Reliability Data Collection and Analysis

EURO

COURSES

A series devoted to the publication of courses and educational seminars organized by the Joint Research Centre Ispra, as part of its education and training program.
Published for the Commission of the European Communities, Directorate-General Telecommunications, Information Industries and Innovation, Scientific and Technical Communications Service.

The EUROCOURSES consist of the following subseries:

- Advanced Scientific Techniques
- Chemical and Environmental Science
- Energy Systems and Technology
- Environmental Impact Assessment
- Health Physics and Radiation Protection
- Computer and Information Science
- Mechanical and Materials Science
- Nuclear Science and Technology
- Reliability and Risk Analysis
- Remote Sensing
- Technological Innovation

RELIABILITY AND RISK ANALYSIS

Volume 3

The publisher will accept continuation orders for this series which may be cancelled at any time and which provide for automatic billing and shipping of each title in the series upon publication. Please write for details.

Reliability Data Collection and Analysis

Edited by

J. Flamm and T. Luisi

Commission of the European Communities,
Joint Research Centre, Institute for Systems Engineering and Informatics,
Ispra, Italy

SPRINGER-SCIENCE+BUSINESS MEDIA, B.V.

Publication arrangements by
Commission of the European Communities
Directorate-General Telecommunications, Information Industries and Innovation,
Scientific and Technical Communication Unit, Luxembourg

EUR 14205

Originally published by Kluwer Academic Publishers in 1992
Softcover reprint of the hardcover 1st edition 1992

ISBN 978-94-010-5075-3
ISBN 978-94-011-2438-6 (eBook)
DOI 10.1007/978-94-011-2438-6

CONTENTS

PREFACE

The ever increasing public demand and the setting-up of national and international legislation on safety assessment of potentially dangerous plants require that a correspondingly increased effort be devoted by regulatory bodies and industrial organisations to collect reliability data in order to produce safety analyses. Reliability data are also needed to assess availability of plants and services and to improve quality of production processes, in particular, to meet the needs of plant operators and/or designers regarding maintenance planning, production availability, etc.

The need for an educational effort in the field of data acquisition and processing has been stressed within the framework of EuReDatA, an association of organisations operating reliability data banks.

This association aims to promote data exchange and pooling of data between organisations and to encourage the adoption of compatible standards and basic definitions for a consistent exchange of reliability data.

Such basic definitions are considered to be essential in order to improve data quality. To cover issues directly linked to the above areas ample space is devoted to the definition of failure events, common cause and human error data, feedback of operational and disturbance data, event data analysis, lifetime distributions, cumulative distribution functions, density functions, Bayesian inference methods, multivariate analysis, fuzzy sets and possibility theory, etc.

Improving the coherence of data entries in the widest possible sense is paramount to the usefulness of such data banks for safety analysts, operators, legislators as much as designers and it is hoped that in this context the present collection of state-of-the-art presentations can stimulate further refinements in the many areas of application.

T. LUISI G. VOLTA

LIST OF CONTRIBUTORS

S. BALESTRERI
CEC, JRC Ispra, Institute of Systems Engineering and Informatics
SER Division, I-21020 Ispra (VA)

G.F. CAMMACK
British Petrolium International Ltd, Britannic House, Moor Lane
London EC2Y 9BU, UK

R.M. COOKE
Dept. of Mathematics and Informatics, Delft University of Technology
P.O. Box 5031, NL-2600 GA Delft, The Netherlands

D. DUBOIS
Inst. de Recherche en Informatique de Toulouse
Université Paul Sabatier
118 route de Narbonne, F-31062 Toulouse Cedex

N. GARNIER
Centre National d'Etudes des Télecommunications
B.P. 40, F-22301 Lannion

L.J.B. KOEHORST
TNO, Div. of Technology for Society, Dept. of Industrial Safety
P.O. Box 342, NL-7300 AH Apeldorn

K. LAAKSO
Technical Research Centre of Finland (VTT/SÄH), Laboratory of Electrical
Engineering and Automation Technology, SF-02150 Espoo, Finland

A. LANNOY
EDF - Group Retour d'Expérience, Dept. REME
25, allée privée, Carrefour Pleyel, F-93206 Saint-Denis Cedex 1

S. LYDERSEN
SINTEF, Division of Safety and Reliability
N-7034 Trondheim, Norway

A. LYYTIKÄINEN
Technical Research Centre of Finland (VTT/SÄH), Laboratory of Electrical
Engineering and Automation Technology, SF-02150 Espoo, Finland

T.R. MOSS
R.M. Consultants Ltd, Suite 7, Hitching Court, Abingdon Business Park
Abingdon, Oxon OX14 1RA, UK

H. PROCACCIA
EDF, Direction des Etudes et Recherches, Dept. REME,
25, allée privée, Carrefour Pleyel, F-93206 Saint-Denis Cedex 1

P. PYY
Technical Research Centre of Finland (VTT/SÄH), Laboratory of Electrical Engineering and Automation Technology, SF-02150 Espoo, Finland

M. RAUSAND
Norwegian Institute of Technology, Division of Machine Design
N-7034 Trondheim, Norway

H.A. SANDTORV
SINTEF, Division of Safety and Reliability
N-7034 Trondheim, Norway

H.J. WINGENDER
NUKEM GmbH, P.O. Box 13 13
D(W)-Alzenau, F.R.G.

PRESENTATION OF EuReDatA

H. PROCACCIA
EDF, Direction des Etudes et Recherches
Département REME
25, allée privée, Carrefour Pleyel
F-93206 Saint-Denis Cedex 1

Preface

EuReDatA is an Association having for goal to facilitate and harmonize the development and operation of reliability, availability or events data banks of its members. In particular:

- to promote data exchange between organizations and to encourage comparison exercices between members,
- to establish a forum for the exchange of data bank operating experience,
- to encourage the adoption of compatible standards and definitions for data, and to establish guides for collecting and analyzing these data,
- to set up agreed methods for data authentification, qualification and validation,
- to promote training and education in the field.

Table of contents

J. Flamm and T. Luisi (eds.), Reliability Data Collection and Analysis, 1–13.

1. History of the Association

The EuReDatA Group was formed in 1973 as a result of discussions at the First European Reliability Data Bank Seminar held in Stockholm. It was first an association of European organizations constituted firstly to solve the problems encountered in setting up and managing reliability data banks. The second objective was the adoption of agreed procedures in certain key areas of activity, in order to form a common language permitting the exchange of data among member banks.

These first objectives have been extended later to availability and event data banks.

The Group was formally constituted as the European Reliability Data Banks Association (EuReDatA) on 5 October 1979 with the support of the Commission of the European Communities, assuring the secretary of the Association.

The founder members of EuReDatA were:

- Commission of the European Communities, Joint Research Centre, Ispra (Italy),
- Centre National d'Etudes des Télécommunications, Lannion (France),
- Det Norske Veritas, Oslo (Norway),
- Electricité de France, Paris (France),
- Ente Nazionale Idrocarburi, Milano (Italy),
- European Space Agency, Paris (France),
- Istituto Elettrotecnico Nazionale «Galileo Ferraris», Torino (Italy),
- TNO, Netherlands Organization for Applied Scientific Research, Apeldoorn (Netherlands),
- United Kingdom Atomic Energy Authority, Warrington (U.K.),
- RM Consultants Limited, Abingdon (U.K.),
- Arne Ullman AB, Saltsjöbaden (Sweden) (founder).

Since its foundation, EuReDatA has grown to its present total of 48 members (Status at end 1989) who collectively renewed the constitutional agreement of the Association (list given in appendix 1). This agreement nominates as Honorary Chairman Arne Ullman, who was the first chairman of the Assembly and greatly contributed to the foundation of EuReDatA.

In the meantime, as shown later, the Association has promoted many significant activities in the field of data collection and analysis, project groups, seminars, conferences and courses.

Still maintaining its autonomy, EuReDatA interacts with ESRA (European Safety and Reliability Association) which is a new initiative promoted by the Commission of the European Communities in order to stimulate and coordinate research, education and activities in the safety and reliability field. One member of the Executive Committee of EuReDatA is one of the members of the Steering Committee of ESRA.

On the other hand, it maintains close relations with ESRRDA (European Safety and Reliability Research and Development Association) depending of ESRA.

To face the need of funding project groups and therefore becoming a more authoritative data suppliers club, the actual trend is to move towards the establishment of EuReDatA as a non-profit organization with siege in Luxembourg, or Brussels, in 1993.

2. Membership

A member of EuReDatA can be any organization in EEC or EFTA countries, private, governmental or other, operating or planning to build and operate a data bank, either of reliability or incident data. A matrix given in appendix 2 gives the partition of the present members.

Since 1988, special provisions have been introduced into the Agreement in order to open the Association to members not belonging to European countries.

Each member commits itself:

- to promote the objectives of the Association and in doing so, adopts agreed definitions and procedures aimed at the exchange of reliability information and experience,
- to demonstrate that the organization has the capability to fulfil the commitments and requirements as stated above.

3. Financing

EuReDatA is a non-profit association which does not require subscription fees until now. Each member covers its own expenses.

Presently, the general secretary is assured by the Joint Research Center of ISPRA.

4. Main Topics of the Constitutional Agreement of EuReDatA

The Association is organized with a **chairman** helped by an **Executive Committee** and a **General Secretary** and an **Assembly** of members.

- The **CHAIRMAN** is elected for at least 2 years by the members of the Association.
  His role is to look at the respect of the Constitutional agreement and to promote topics regarding the development, the operation, the exchange, the analyses, the standardization and the quality of data banks.
  On the other side, he has to encourage creation of specific project groups, to support seminars, conferences or courses.
- The **ASSEMBLY**, composed of representatives of the members, establishes the policy of the Association, in particular it has to identify those data banks topics requiring investigation, to encourage the execution of these investigations, through setting up Project Groups, to resolve individual problems arising from such joint ventures, and to organize technical and scientific symposia which may be open to the public.
  The Assembly elects the Chairman of the Association, and votes for the admission of a new member.
  The Assembly presently appoints a General Secretary on the nomination of the Commission of the European Communities. The General Secretary is currently resident at the Ispra Establishment of the Joint Research Centre of the Commission of the European Communities.
- The **EXECUTIVE COMMITTEE** assists the Assembly in the preparation of its decisions. In particular, it coordinates the Project Group activities, investigates and reports to the Assembly on new membership applications, considers policies and actions for decision by the Assembly, and approves external publications of the Association.
  Members of the Executive Committee are:
  . the Chairman of the Association

  . the preceding Chairman of the Association
  . the General Secretary of the Association
  . three members of the Association elected every 2 years by the Assembly.

  The Executive Committee meets at least four times a year.

- The **GENERAL SECRETARY**. The function of the General Secretary is to ensure the satisfactory running of the Association by supporting and servicing various bodies of the Association. The appointment of the General Secretary, nominated by the Commission of the European Communities, requires the majority of the votes of the members of the Assembly.
  The General Secretary is assisted by a Secretariat for the execution of its tasks.
- **PROJECT GROUPS**. The main objective of the Project Groups is to execute specific tasks or programmes of interest for the aims of the Association. The work programme of each Project Group is defined by the Assembly under proposal of the Executive Committee. Each Project Group is coordinated by a Group Leader which reports to the Assembly on the results of the work carried out by the ad-hoc Project Group. They are open to collaboration with external experts.

5. Operation of EuReDatA

The Association proposes to keep a membership file, containing information about the member data banks judged to be of possible interest to others. This file is available to all members.

The general experience of keeping data files and of the acquisition, evaluation and use of data from different sources can be exchanged freely between members, directly or at meetings and seminars.

The specific information about reliability parameters is of a different character and not freely published. The exchange or pooling of data and possibly acquiring of data by one or more members from a fellow member is to be agreed upon directly by the members concerned. The Association shall not be directly concerned with the conditions for such agreements.

If data have to be disclosed to fellow members in the project group work, these data remain strictly the property of their owners. The text in the constitutional Agreement is formulated to cover this situation.

The project groups will not duplicate work done by ISO, IEC, EOQC or others working with reliability definitions and standards, but will base their work on the internationally achieved results.

One example of reference classification concerning valves reliability (extracted from project report 1) is given in appendix 3. The same reference classification has been made for:

- emergency diesel generator sets,
- electric motors,
- actuators:
  . electro-mechanical,
  . hydraulic,
  . pneumatic.
- electronic components.

A list of publications available at the EuReDatA Secretary is given in appendix 4, and a typical data bank form is given in appendix 5.

Appendix 1
EuReDatA Members/Representatives (1990)

Denmark
- Danish Engineering Academy
  Mr. Lars Rimestad

Finland
- Imatran Voina Oy (IVO)
  Mr. Pekka Louko
- Industrial Power Company Ltd. (TVO)
  Mr. Risto Himanen
- Technical Research Centre (VTT)
  Mr. Antii Lyytikäinen

France
- Electricité de France (EDF)
  Mr. H. Procaccia
- Institut Français du Pétrole
  Mr. A. Bertrand
  Mr. R. Grollier Baron
- Renault Automation
  Mr. B. Dupoux
- TOTAL - CFP
  Dr. J.L. Dumas

F.R. Germany
- Interatom
  Mr. J. Blombach
- NUKEM GmbH
  Dr H.J. Wingender

Ireland
- Electricity Supply Board (ESB)
  Mr. Vincent Ryan

Italy
- EDRA
  Mr. M. Melis
- ENEA
  Dr. C.A. Clarotti
- ICARO Srl
  Mr. Giancarlo Bello
- Donegani Anticorrosione
  Dr. Carlo A. Farina
- ITALTEL S.I.T.
  Mr. G. Turconi
- TECSA Srl
  Mr. C. Fiorentini
  Mr. A. Lancia
- TEMA/ENI
  Mrs. V. Colombari

The Netherlands
- N.V. KEMA
  Mr. R.W. van Otterloo
  Mr. J.P. van Gestel
- TNO
  Mr. P. Bockholts
  Mr. L. Koehorst

Norway
- Det norske Veritas
  Mr. Morten Sorum
- Norsk Hydro
  Mr. T. Leinum
- SIKTEC A/S
  Mr. Jan Erik Vinnem
- SINTEF
  Mr. Stian Lydersen
- STATKRAFT
  Mr. Ole Gjerde
- STATOIL
  Mr. H.J. Grundt

Spain
- TEMA S.A.
  Mr. Alberto Tasias
  Mr. J. Renau

Sweden
- AB VOLVO
  Mr. S. Vikman
- Ericsson Radar Electronics AB
  Mr. Markling
- VATTENFALL
  Mr. J. Silva

Switzerland
- Motor Colombus Consulting Eng. Inc.
  Dr. V. Ionescu

United Kingdom
- Advanced Mechanics Engineering
  Mr. C.P. Ellinas
- BP Int. Ltd.
  Mr. G.F. Cammack
- British Nuclear Fuels plc
  Mr. W.J. Bowers
- CEGB
  Mr. R.H. Pope
- GEC Marconic Research
  Mr. D.J. Lawson
- Health & Safety Executive (HSE)
  Dr. F.K. Groszmann
- Int. Computers Ltd.
  Mr. M.R. Drury
- Loughborough Univ. of Technology
  Prof. D.S. Campbell
- Lucas Aerospace Ltd
  Mr. P. Whittle
- Lucas Rail Products
  Mr. I.I. Barody
- NCSR - UKAEA
  Dr. N.J. Holloway
- RM Consultants Ltd.
  Mr. T.R. Moss
- Nottingham Polytechnic
  Prof. A. Bendell
- University of Bradford
  Dr. A.Z. Keller
- Yard Ltd. Consulting Engineers
  Mr. I.F. MacDonald
- Electrowatt Eng. Services Ltd
  Mr. G. Hensley

Commission of the European Communities
- JRC Ispra
  Mr. G. Mancini

Appendix 2.1
EuReDatA Matrix

	Data Supplier Classification			
		Authority/ Certification Agency	**Consultant**	**Manufacturer**
	Chem./Petro. Offshore	-Health & Safety Executive (HSE) (UK)	- TEMA/ENI (I) - TEMA S.A. (E) - SIKTEC A/S (N) - TECSA Srl (I)	- BP Int. (UK) - Norsk Hydro (N) - STATOIL (N) - NUKEM GmbH (D)
	Electrical			
	Electronic		- ICARO Srl (I) - TECSA Srl (I)	- Ericsson Radar Electronics AB (S) - Int. Computers Ltd (ICL) UK) - ITALTEL S.I.T. (I) - GEC Marconi Res. Centre (UK)
	Mechanical	- Det norske Veritas (N)	- N.V. KEMA (NL) - R.M. Consultants Ltd. (UK) - Yard Ltd Consulting Eng. (UK)	- NUKEM GmbH (D)
	Nuclear	- ENEA (I) - Health & Safety Exec. (HSE) (UK)	- EDRA (I) - Motor Colombus C.E. Inc. (CH) - N.V. KEMA (NL) - R.M. Consultants Ltd (UK)	- British Nucl. Fuels (UK) - Interatom (D) - NUKEM GmbH (D)
	Car/Vehicle Railways Aircraft/Space			- AB VOLVO (S) - Lucas Rail Products (UK) - Renault Automation (F)

Appendix 2.2

	Data Supplier Classification			
		Research Institute	University	Utility
	Chem./Petro. Offshore	- Inst. Français du Pétrole (F) - Ist. Donegani Spa. Montedison (I) - TNO (NL)		- TOTAL, CFP
	Electrical	- SINTEF (N)		- CEGB (UK) - EDF (F) - Imatran VOIMA OY (IVO) (SF) - STATKRAFT (N) - VATTENFALL (S) - TOTAL, CFP
	Electronic	- Techn. Res. Centre of Finland (VTT) (SF)	- Danish Eng. Academy (DK) - Loughborough Univ. of Techn. (LUT) (UK)	
	Mechanical	- JRC Ispra (CEC)	- Trent Polytechnic (UK) - Univ. of Bradford (UK)	- CEGB (UK) - EdF (F) - VATTENFAL (S) - TOTAL, CFP - Imatran VOIMA OY (IVO) (SF)
	Nuclear	- JRC Ispra (CEC) - NCSR-UKAEA (UK) Techn. Res. Centre of Finland (VTT) (SF)		- Industrial Power Comp. Ltd (SF) - N.V. KEMA (NL) - VATTENFALL (S) - EdF (F) - Imatran VOIMA OY (IVO) (SF)
	Car/Vehicle Railways Aircraft/Space			

Appendix 3.1
Mechanical valves reference classification

- VALV
  - Design Related
    - 01 - Type
    - 02 - Function/Application
    - 03 - Actuation
    - Capacity Performance
      - 04 - Size (SZ) (Nominal diameter)
      - 05 - Design Pressure (PR)
      - 06 - Design Temperature (TE)
    - Construction features
      - Materials
        - 07 - Body Material (MA)
        - 08 - Seat Material (MA)
        - 09 - Disc Material (MA)
      - 10 - Body Construction Type (MP)
      - 11 - Seat Type (C0)
      - Sealing
        - 12 - Valve externally (SA; SB; SC)
        - 13 - Valve Internally (SA; SB; SC)
    - 14 - Safety Class/Standards
  - Use/Application Related
    - Process Related
      - 15 - Process Pressure (PR)
      - 16 - Process Temperature (TE)
      - 17 - Medium Handled (MH)
    - Environment Related
      - 18 - Type of Industry (EI)
      - 19 - Vibrations (EV)
      - 20 - (Environmental) Temperature (ET)
      - 21 - Radiation (EV)
      - 22 - Type of Installation (EL)
      - 23 - Position Relative to Sea-level (EA)
      - 24 - Climate (EC)
      - 25 - Humidity (EH)
      - 26 - (Environmental) Influences (EE)
      - 27 - (Environmental) Pressure (EP)
    - 28 - Maintenance Related (MS)
    - 29 - Duty Related (M0)

Appendix 3.2
Descriptors unique to mechanical valves

Category 01: Type	*Code*
Ball	10
Butterfly	20
Check N.O.C.	30
Check, swing	31
Check, lift	32
Cylinder (piston & ports)	40
Diaphragm	50
Gate (sluice, wedge, split wedge)	60
Globe N.O.C.	70
Globe, single seat	71
Globe, single seat, cage trim	72
Globe, double seat	73
Needle	80
Plug	90
Poppet	A0
Sleeve	B0
Other	ZZ

Category 02: Function/Application	
Bleed	10
Bypass	20
Control/regulation	30
Dump	40
Exhaust	50
Isolation/stop	60
Metering	70
Non-return/check	80
Pilot	90
Pressure reducing	A0
Relief/safety	B0
Selector (multiport valve)	C0
Vent	D0
Other	ZZ

Category 03: Actuation	
Differential pressure/spring	10
Electric motor/servo	20
Float	30
Hydraulic	40
Pneumatic	50
Mechanical transmission	60
Solenoid	70
Thermal	80
Manual	90
Other	ZZ

Appendix 3.3
Mechanical valves (VALV)

Boundary definition

Component boundary is identified by its interfaces with the coupling/connections to the process system. The valve actuator and associated mechanisms are considered to be part of the mechanical valve.

When power actuators are utilized, the actuator should be identified according to the item identification for hydraulic, electric and pneumatic actuators.

Appendix 4
Publications by EuReDatA
(Status 1990)

Proceedings of EuReDatA Seminars

1. VTT Symposium 32 «Reliability Data Collection and Validation", October 1982, Helsinki, Finland Government Printing Centre, P.O. Box 156, SF-00101 Helsinki 10.

2. Symposium on Materials Reliability, October 1983, Baden, Switzerland. Published by Butterworths.

3. Reliability of Rotating Machinery, April 1984.

4. Reliability of Automatic Fire and Gas Detector Systems, July 1984.

5. Use of Reliability Data in Major Hazard Assessment, October 1984.

6. International Cooperation in Reliability and Safety Data and their use for Large Industrial Systems, April 1985.

7. Accident Data Banks, october 1985.

8. Fire Data Analysis and Reliability of Fire Fighting Equipments, October 1986.

9. Case Studies on Availability Assessment, April 1987.

10. Reliability Data Acquisition and Utilization in Industrial Electronics, October 1987 (not yet available).

11. The Use of RAM-Data in the Decision Making Process, September 1988.

12. EuReDatA Seminar on Maintenance, January 1989.

13. EuReDatA benchmark exercice on reliability data analysis, April 1990.

Ispra Course on Reliability Data Bases,

D. Reidel Publ. Co., Dordrecht (NL), 1987

Eurocourse: Reliability Data collection and analysis
KLUVER Academic Publisher (1990).

Proceedings of EuReDatA Conferences on

«Reliability Data Collection and Use in Risk and Availability Assessment»

1. Stockholm, November 1973 (FOA/FTL A 16:41).

2. Stockholm, April 1977 (FOA/FTL A 16:69); *both available from National Defence Research Institute Library, P.O. Box 1165, S-581 11 Linköping.*

3. Bradford, April 1980: *available from UKAEA Course Conference Organiser, Wigshow Lane, Culcheth, Warrington, WA34NE, U.K.*

4. Venice, March 1983: *available as microfiches from NUKEM, Dr. H.J. Wingender, Postfach 1313, D-8755 Alzenau, FRG.*

5. Heidelberg, April 1986: *published by Springer Verlag, N.Y., Heidelberg, Berlin.*

6. Siena, March 1989: *published by Springer Verlag, N.Y., Heidelberg, Berlin.*

Project Reports

No. 1 Reference classification concerning components reliability.

No. 2 Proposal for a minimum set of parameters in order to exchange reliability data on electronic components.

No. 3 Guide to reliability data collection, validation, storage.

No. 4 Materials reliability.

No. 5 Reference classification concerning Automatic Fire and Gas Detection Systems (AFGDS) (*in preparation*).

No. 6 Characteristics of Incident Databases for Risk Assessment (*in preparation*).

EuReDatA Chairman (90)

Electricité de France	Mr. H. PROCACCIA
Direction des Etudes et Recherches	Tél. : 331 49 22 89 38
25 Allée Privée	Telefax : 331 49 22 88 24
93206 St Denis (France)	Telex: 231889FEDFRSD

EuReDatA General Secretariat

Commission of the European Communities	Mr. T. Luisi.
Joint Research Centre - Ispra Site	Tel. + 332-789471
Systems Engineering and Reliability	Telex : 380042/38995 EUR I
Division	Telefax + 39-332-789001
I-21020 Ispra (VA) (Italy)	and + 39-332-789472

Appendix 5
EuReDatA Data Bank Form

DATA BANK FORM

DATA BANK CODE	NAME : SRDF/RPDF		COUNTRY : FRANCE	RESPONSIBLE : A. LANNOY PHONE 33.1.49228923 FAX : 33.1.49228824		
	INITIATION DATE : 1978		STATUS : in development [ ]	in operation [X]	stand by [ ]	projected [ ]
TYPE	STATISTIC [ ]	EVENT [ ]	RELIABILITY [X]	MAINTENANCE [ ]	CONTROL [ ]	OTHERS [ ]
TYPE OF DATA	RAWDATA [X]	ELABORATED [X]	PARAMETRIC [X]	KEYWORD [ ]		CODED [X] NON CODED [ ]
SOURCES	MANUFACTURER [ ] INDUSTRY [ ]	UTILITY [X]	AUTHORITY [ ]	PUBLIC [ ]	LITTERATURE [ ]	OTHERS [ ]
TYPE AND NUMBER OF ITEMS/TYPE	ELECTRONICAL	ELECTRICAL	MECHANICAL	SOFTWARE	ACCIDENT	OTHERS
		X 20 000	X 30 000			
SAMPLING 1-1-90 NB OF FORMS	DESCRIPTIVE	OPERATING	FAILURES	EVENTS	ACCIDENT	PARAMETRIC DATA
	X 25 000	X 200 000	X 25 000			X ON DEMAND
FAILURE/ACCIDENT CHARACTERISTICS	IN OPERATION [X]	PROGRESSIVE/ PARTIAL	CRITICAL/ COMPLETE	CONSEQUENCES/ SUBCOMPONENT	MODE	CAUSE
	DURING STANDBY [X]	X	X	X	X	X
INFORMATIC SUPPORT	TYPE OF COMPUTER	DBMS	SOFTWARE	MANUAL		OTHERS
	IBM 3090	DB2	SAS/SADE			
ACCESS/ COST	FREE	RESTRICTED	CONFIDENTIAL	WITH CHARGE	CUMULATIVE COST / ANNUAL COST	OTHERS
			X	(FUTURE)	100 10^6 FF / 26 10^6 FF	

NEEDS AND USE OF DATA COLLECTION AND ANALYSIS

H.J. WINGENDER
NUKEM GmbH
P.O. Box 13 13
D-8755 ALZENAU, F.R.G.

1. Introduction

1.1 Expectations

First of all, I wish to thank EuReDatA and the organisers of the course for honouring me with the privilege of reading a general introduction to the course. As is usually the case with honours and privileges, they are accepted with great pleasure but also a little trepidation as to whether the expectations that go along with them can be met.

I have been thinking therefore, about what those expectations might be. Taking into consideration the objectives of the European Reliability Data Bank Association EuReDatA and those of the course, I finally concluded that I am expected to encourage as much discussion and communication as possible amongst all participants right from the beginning. Consequently, I understand introduction verbally: guide us over the threshold and through the entrance and establish an open collaborative attitude which should be maintained throughout the course. The next two subsections are intended to show how this will be achieved and then, hopefully, in the paragraphs dealing with needs and use of data collection, needs and use of data analysis, followed by warnings and conclusions, those expectations I mentioned earlier will be met.

1.2 General Ideas about the Course

As indicated in the announcement, the course is intended for scientists and engineers active in reliability engineering, and in particular for those planning the installation and the use of reliability data banks. The course shall provide the delegates with the experience - or rather

J. Flamm and T. Luisi (eds.), Reliability Data Collection and Analysis, 15–43.

with some of the experience accumulated at several of the member organisations of EuReDatA. Furthermore, I am expecting extensive information exchange amongst the participants during the course and the establishment of mutual communication links between the participants for the time after the course. There is also a fair chance that the lecturers will learn more from the trainees' experience because of the sheer ratio of people on both sides.

In order to make the course a success for all the participants, i.e. to convey as much valuable, practice-based information as possible from the lecturers to the auditory and vice versa, the lecturers are supposed to give sufficient time for discussions and to make extensive use of examples and demonstrations. To this end the course is properly structured. It starts with the basic definitions and requirements, proceeds then to how it all works: the collection and the processing of data, the implications following from data uncertainties and how to cope with them, the structure and operation of data banks and how to use them, and finally the data analysis.

1.3 Specific Ideas about this Lecture

The purpose of such an introductory lecture cannot be to give a condensed version of the complete course, because that would be technically impossible and eventually obsolete in view of the objectives of the course. It is however equally inevitable that some of the subjects or items exhaustively treated in the lectures to follow are touched here.

Considering the objectives of the course, I decided to focus this general overview of the needs and use of data collection and analysis on two particular aspects:

- rough outlining of a couple of questions one should bear in mind and may or may not ask during the discussions,
- coarse indications of a couple of difficulties you may have faced or may still face during your work as reliability engineers.

As mentioned in the previous paragraph the lectures show how the data business works. However, being engineers we all know that there is more to be considered than how something simply works. There are, for instance, the questions of how to get it to work, and why it does work at all or in a particular way. Furthermore, as reliability engineers we are supposed to ask, why, how and how often a system will fail, and with what consequences, and what can be done about it.

I do not claim that you will find all the answers or solutions from this course or anywhere else. I even doubt that all of them are currently at our disposal. But I am of the opinion that all those questions and difficulties are important for the work of a reliability engineer - even the seemingly far fetched ones - and I am sure that there are more questions and difficulties behind those mentioned of which I am not aware and not capable of formulating at present.

From the many definitions of what an expert is I am inclined to the one which states that an expert is aware of the major mistakes possible in his/her subject and of the best ways of avoiding them. Thus, I hope that this lecture will help you to extract exactly this type of information from the other lecturers and will eventually lead some of you into pursuing those points remaining unanswered and unsolved for the time being.

2. The Needs and Use of Data Collection

2.1 Importance of Reliability Data

Recently it has become a custom in Europe to choose the metaphor of a house when one is attempting to explain a new idea or to put forward an unfamiliar view of a subject. Reliability engineering in all its forms and uses resembles a house in that it is purposeful, can be of complex design and structure, and should have a sound foundation, of which one usually cannot recognise very much even when looking from inside. Obviously this barely recognisable foundation of reliability is data. Hence, it requires the most careful consideration and cultivation - or in our terms: maintenance and quality control - because otherwise it can cause the complete break down of the availability of an indispensable piece of the system it is supposed to support and consequently of the system in total. A striking example of this kind is the poor foundation of one pillar of the Autobahn bridge crossing the river Inn near Kufstein in Austria. Because of this failure, the whole ground based traffic system crossing the Alps via the Brenner route broke down this summer and will remain in this state for two or three years with respect to car traffic. Train traffic on the main line crossing below the bridge has been re-established in the meanwhile.

What are we talking about, when we refer to reliability data, and why is it so difficult to obtain?

In general terms, reliability data is a piece of quantified experience which - in principle - can be used for the quantitative judgement of the behaviour of a technical system in existence or being planned. The key term is experience because there cannot be any judgement at all, unless it is based upon experience. A technical system can be a mere piece of hard ware or a man-machine system. More complicated is the term "quantified". It comprises the design information of a particular type of component, its operational environment, its mode of operation and its failure behaviour (Table 1).

According to my understanding of quantified data, the term comprises at least the information compiled in table 1 under the headings "basic data" and "derived data", although it very often happens that only a subset is really needed for a particular investigation. The complete information is, however, indispensable in all those cases which aim at the improvement of system or plant reliability, the establishment of maintenance strategies, the planning of back fitting and plant life extension measures and the planning of new systems.

As can be seen from table 1, which makes use of information compiled in reference /1/, there is a distinction to be made between basic data and derived data. It is usually the derived data which reliability engineers are in search of. Because of the rarity of event data or field data collection and analysis at their own establishments, they are frequently forced to make use of published information such as in /2, 3, 4/. Such information, although very valuable in the proper area, is often outdated, incomplete or not applicable for the particular exercise in question. As a consequence engineers have to combine inconsistent data and make use of expert opinion thus ending up with a database of questionable quality and all its consequences stated above.

Facing this difficulty themselves, members of EuReDatA turned their awareness into action, attempting to enable and to facilitate the exchange of reliability data. The first approach was the establishment of a reference classification scheme of component reliability /5/ followed by a proposal for a limited set of parameters for the exchange of reliability data on electronic components /6/. A third attempt for materials reliability data /7/ states in this respect: "With regard to data banking and the exchange of reliability data in this particular field, it is concluded that it is not possible to specify a minimum set of items to permit the ready exchange of data."

Table 1: Reliability data, quantified experience

Data sources:
- Expert opinion
- Laboratory testing
- Published information
- Event data collection
- Field data collection

Basic data (of a component such as valve or pump)
- Engineering description
- Boundary conditions
- Design parameters
- Item identification
- Installation environment
- Operating parameters
- Operating regime or mode
- Maintenance and testing regime
- Event history information
- Failed part information
- Repair information
- Failure cause information
- Failure consequence information

Derived data (usually from a set of components)

* statistical or reliability parameters
  - failure rate, repair rate, availability
  - failure probability
  - mean time to failure
  - mean time between failures
  - mean time to repair
  - probability distributions
  - parameter uncertainties
* non stochastic information
  - data contamination
  - data dependency
  - dependency patterns
  - pattern diagrams
  - deviation from randomness

Having described the understanding of the term data, it should have become clear that the meaning is of some complexity and comprises several levels of detail. One always should be aware of this fact during the course and try to find out what the actual meaning is when the term is used.

The complexity of data and, in particular, of the information required to establish a complete data set provides a first clue to the answer to the second part of the initial question: why is it that difficult to obtain reliability data. It is understandable that it is not easy and is certainly expensive to install a comprehensive data collection and data evaluation scheme. Such a scheme affects the people operating a plant, puts an extra work load upon them, and is not easily explained to them as something which supports their work. Because of these technical, financial and psychological obstacles, there are not so many data banks as one may expect, taking into account their obvious advantages. Due to the same reasons, operating data banks and their inventories are thought of as highly valuable property which one does not like to share light-heartedly with others, including possible competitors.

Organisations making extensive use of their reliability data collection systems experience the direct feed back of information and the also advantages achieved for the operation of their plants and consequently for their products on the market. This experience adversely affects their preparedness to exchange data. Quoting EuReDatA again /7/ on data exchange: "The problem is exacerbated as a result of proprietary and confidentiality considerations."

Taking all these difficulties into consideration, i.e. the complexity of data, the variety of data collection systems, the reluctance of data owners, the lack of unified and widely used data classification systems - they are often deliberately not used because they might reduce the protection of the confidential information - it is rather obvious, why it is difficult to obtain data.

The more puzzling are the facts which demonstrate a completely opposite attitude and which force questions like: why does EuReDatA work at all, and how has an Offshore Reliability Data project become so successful?

2.2 The Need of Data Collection

The need to collect reliability data in a proper way has been evidenced in the foregoing paragraph. There is always the question, however, of whether or not it is really necessary to collect data for a particular exercise in reliability engineering. Nobody is eager to spent money for and to put effort into a task which could be done cheaply and easily. In addition there are those areas and reliability assessment tasks which deal with completely new systems and first-of-its-kind equipment, for which no operating experience exists.

It may be justified to question whether these problems really need extensive collection and analysis of basic data. From my own experience I have concluded that the particular method of data collection needed, the type of data needed and the degree of data consistency needed are essentially governed by the purpose for which the data is to be used and can be decided upon on this basis.

Our company received a contract to assess the reliability and availability of a first-of-its-kind machine for the charging of high level active wastes into the bore hole positions in a repository for radioactive wastes (Figure 1). Many of the components are widely used in other systems, others are of entirely new design. There is mechanical, hydraulic, electrical and electronic equipment in the system (Figure 2). It contains a complex network of inter-

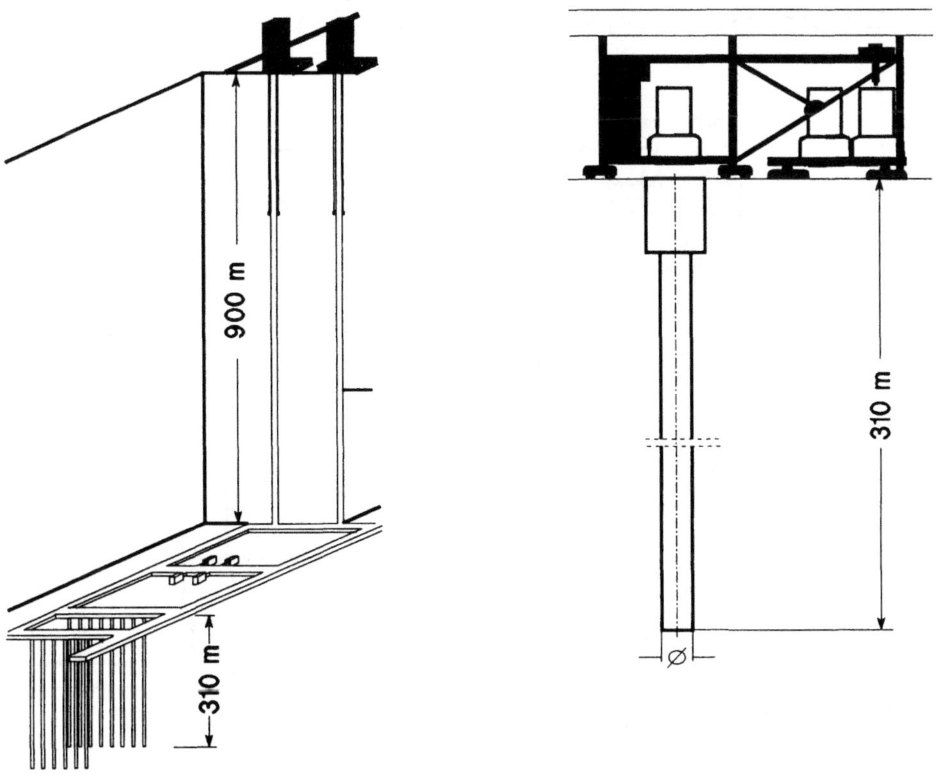

Figure 1: Radioactive waste repository and high level waste emplacement bore hole

locks. It is also supposed to operate highly reliably in a hostile environment containing radiation, rock salt dust, some humidity forming corrosives with the salt dust and at temperatures of around 50°C. A data source properly taking into account all these factors was not at our disposal. We put together - out of necessity - data from the OREDA Handbook /2/, from the Systems Reliability Service of the UKAEA and from our own data banks. It was all derived data, although we made sure that they came from hostile environments. Without detailing the painstaking investigations we performed concerning the machine and the data, in order to find the obvious flaws and eliminate apparent inconsistencies, we came up with a result indicating that the requirements are probably met (Figure 3). We did not stop there, however, but convinced the customer, that a meticulous field test of the machine is necessary and that this test simultaneously should become a data collection exercise, in order to get the proper feed back for system backfitting. We are now finishing the test programme, which will be self controlled in so far as the data itself is concerned - i.e. the failure rates, repair rates etc. derived from the sys-

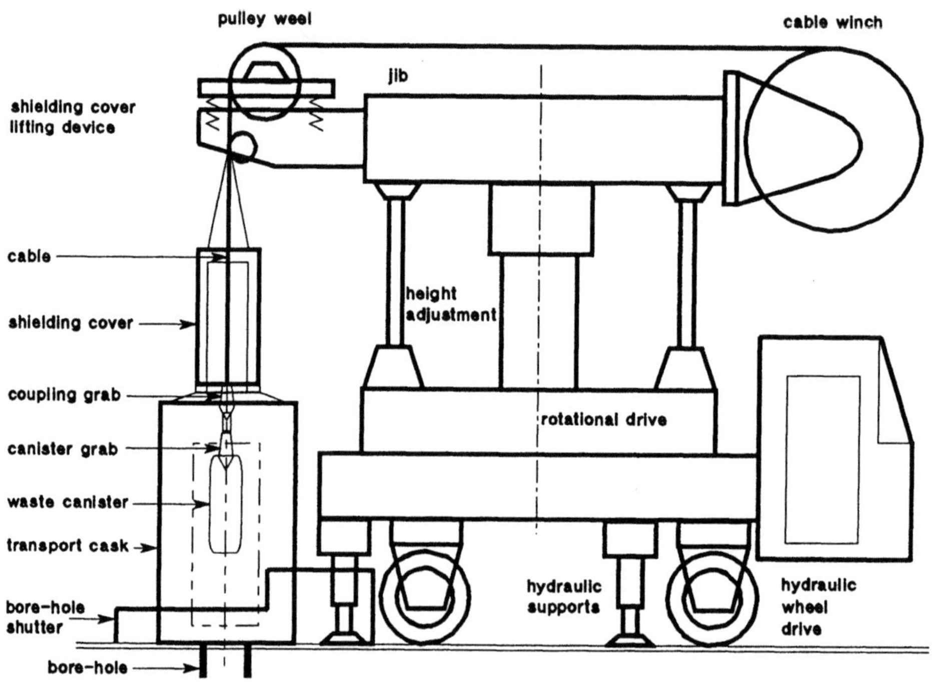

Figure 2: High level waste emplacement system

tem and from the component behaviour and processed through the fault trees are used to set or reset the test runs and the test frequencies, until sound conclusions can be drawn.

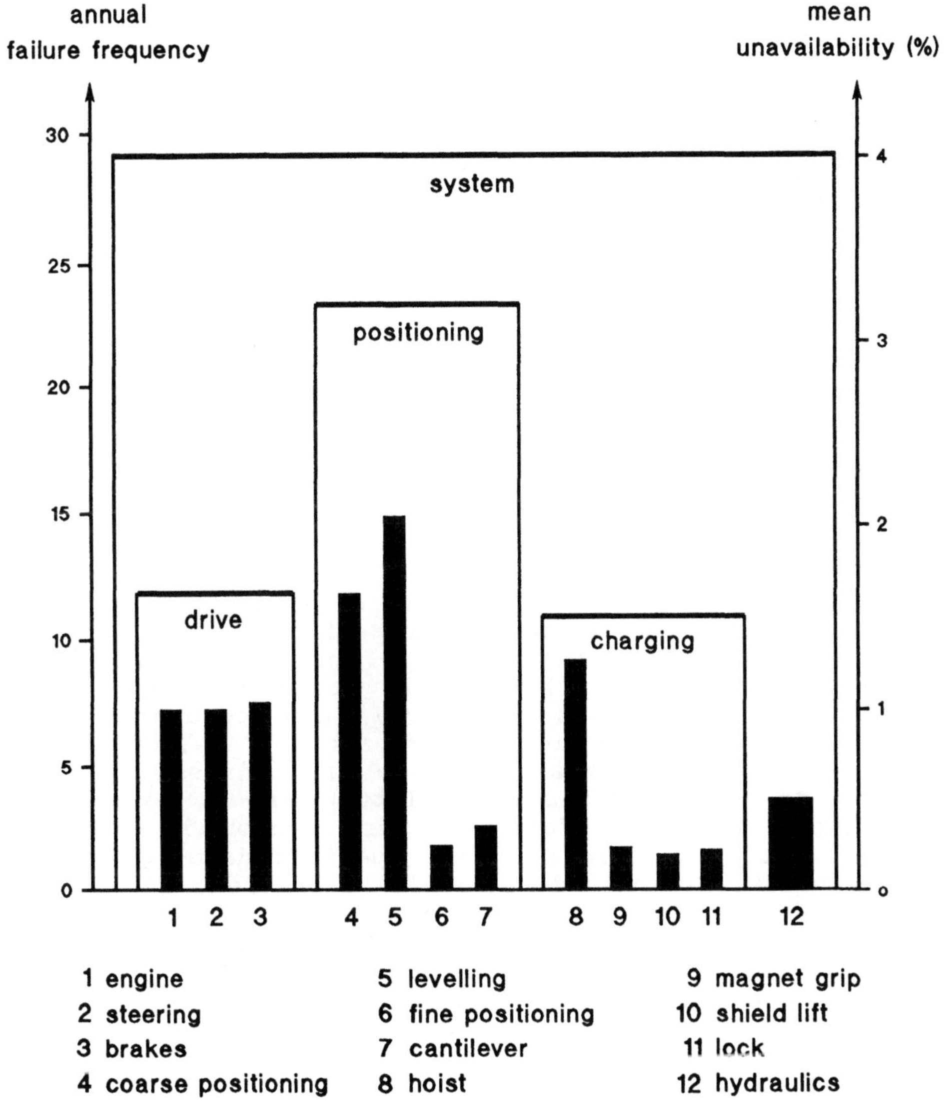

Figure 3: Emplacement system; results of component and system availability assessment

Thus, it may be correct practice to rely upon a poor data base, as long as the conclusions drawn from the result are appropriately judged and used, and are not overvalued. This should however never result in preventing a proper data collection exercise as soon as this is possible.

Another example from my personal experience concerns to probabilistic risk assessment of nuclear power plants. The purpose of the task is self explanatory and obviously quite sensitive. Nevertheless, even semi-official guidelines recommend the use of so-called generic data if none better is available. The recommendation is reasonable in its logic but dangerous in its psychology. It is reasonable in that generic data comprises data collected at plants of a design similar to that for which they will be used. It is also reasonable in that this may be the best data available in the case of no data collection at the plant in question and also reasonable in that generic data are actually available.

The recommendation is however dangerous in that it may lead to the implication that the results of PRA using generic data and a PRA using plant specific data are of equal value. The recommendation is also dangerous in that it may reinforce the reluctance to install a data collection system.

It is sometimes said as a means of comparison that no company exists which uses the data of its nearest competitor - i.e. the most similar one - for the preparation of its own balance sheet, because it could not obtain this generic data, because it is illegal, and because it would give an entirely wrong picture. The argument is, of course, that you could not compare the use of business data with reliability data, because the latter is of far grater uncertainty, and that for this very reason it does not matter, whether generic or plant specific data is used in PRA. Actually, I cannot decide, if the argument is correct. However, I doubt its validity because of experience from practice: If generic data was sufficient, it would be irrational to keep plant specific data a confidential company property, which many companies do, and it would be reasonable to assess the maintenance strategy of a company by means of generic data, which none company does. If it is important to use plant specific experience for maintenance purposes, then it will be inevitable to do the same for safety purposes.

2.3 The Use of Data Collection

One may expect here a compilation of industries using data collection for particular purposes, or of the various purposes for which data collection is a prerequisite. This might be interesting, but it is not intended in this paper.

Whatever task basic data collection is used for, the results will be flawed unless the data is of good quality. Basic data is often referred to as raw data, a proper characterisation, as raw data is like a clutch of raw eggs; both need careful handling und processing. Eggs and data are the carriers of information from which all other conclusions develop: birds if properly reared, otherwise a terrible mess.

Thus, for this lecture I interpret the "use" of data collection as meaning the way it should be done in order to establish the necessarily "clean, uncontaminated and healthy" data base.

For this purpose, some precautionary measures are to be taken right from the beginning:

. The data collection department is to be properly placed in the hierarchy of the company and it is to be equipped with appropriate tools, qualified staff and sufficient authority.

. The employees completing the data forms in the plant are to be properly trained, provided with forms which are easy to handle and motivated to do the job with care.

. The data collection procedure must run continuously. Periodical collection is likely to result in incomplete data bases.

. The procedure for inputting data into the data bank must be so arranged that mistakes are, as far as possible, avoided. The input procedure should be program-controlled in such a way, that data inconsistencies, incomplete data and non-plausible data are indicated on the display and are rejected from permanent storage until correction or confirmation. A printout of the rejected information should be automatically provided and conveyed to the data quality control office, which should take proper measures to ensure immediate examination with the person who recorded the information. Immediate action is vital, because the quality of the examination depends on the memory of the people.

At this point of the procedure the organisation and the structure of the data bank is of great importance. Nevertheless, I refrain from discussing the subject, as it will be extensively treated in one of the subsequent lectures.

Just one remark: when the data are accepted for permanent storage, they are looked at automatically by codes which determine how the data fit into the previous history of the component. It is checked whether this history - and the new data - are compatible, i.e. as expected for the component, and if not, a question mark over this new data is raised. One can argue that this kind of operation is already data analysis. In former times, when computer capacity was smaller and programme languages were less capable, it certainly was analysis. However, more and more of what is analysis today will become data bank operation tomorrow.

Now, having the data in the bank one faces the question, how long should it be kept there? I have no idea at all.

3. The Needs and Use of Data Analysis

3.1 Data Analysis

As said at the beginning of paragraph 2, reliability data is a piece of quantified experience. It has always astonished me how human beings can learn from experience, how information from the past is condensed, interlinked, processed, applied to the present and the future and transferred to other situations, even when these situations are not apparently similar to those from which the experience was drawn.

Data analysis is one link of the chain conveying experience to application. The nature of the data analysis is consequently determined by the particular purpose for which the experience is to be used. Although there is a seemingly unlimited number of imaginable purposes, the number of data analysis methods is currently finite. One of the first steps of data analysis is the determination of frequencies of events. Figure 4 shows an example: pump failures allocated to the sub-group concerned. There are 13 sub-groups within this component class. In the data bank from which this, particular example stems, the sub-groups are broken down further. The "Seals group", for instance, consists of 13 parts due to the different types of pumps und due to the fact that several types of seals are used in a single pump. Just as reminder of the need to keep the data "clean", 67 types of failures are to be allocated for the component class pumps, and there are also other component classes at the plant.

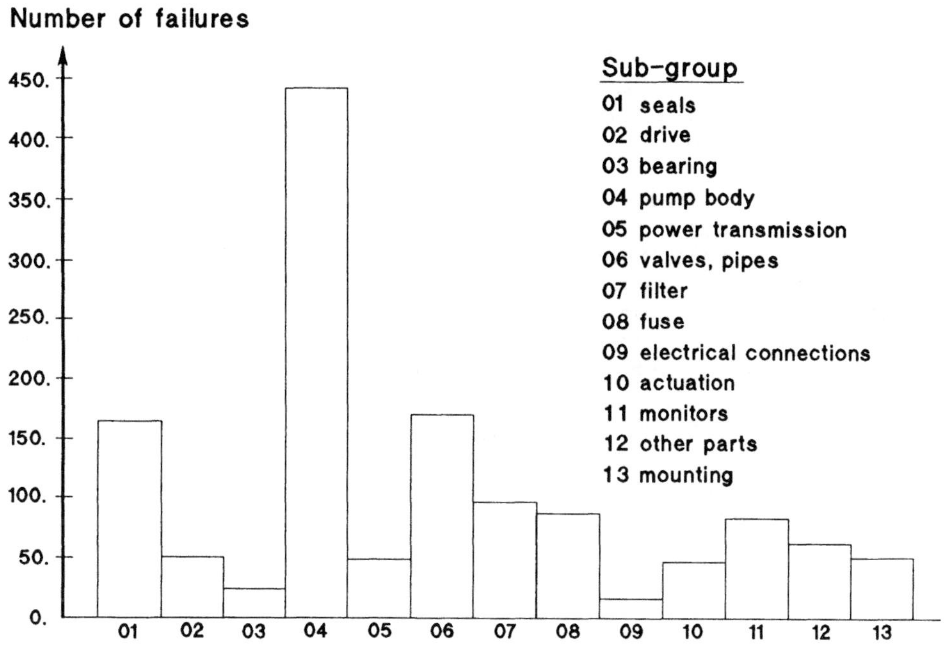

Figure 4: Number of pump failures, determined per sub-group (139 pumps, all types, all modes, 5 years of operation)

Such an analysis tells us that there were roughly 1300 failures during five year operation of this particular population of pumps in their particular modes of operation in this particular plant. This can be extended to repair costs, which is an important piece of information. It does not tell us anything about the development of failures per year during these five years, about failure rates, repair rates, failure probabilities, etc. However, it tells us something else, which is that we do not question the validity of the information. Its derivation is a simple, straight forward accumulation of data.

If however I stated that this would tell us a failure rate of 1300 / (5 x 139) = 2 failures per pump and year, I would be in trouble. You would accuse me rightly, that I based this "analysis" on an inhomogeneous population of items and that I had not shown the exponential distribution of failure probability with time.

What happened between these two very simple and straight forward steps of analysis? It is that the second step has been proven useless by experience, which has led to a very complicated and intriguing concept of what is meant by the term failure rate. First of all, a given failure rate is only applicable to a homogeneous population of items; I do not explain here what that is. Second, it requires the knowlegde of the probability function for failure occurrence versus time. I'll come to that later.

In consequence, the first (and in my opinion most important) task of data analysis is to find out and to validate homogeneous populations of items. As it turns out, that is not always easy, because a population homogeneous in one aspect (parameter) may be inhomogeneous in another aspect. We had a benchmark exercise on data analysis in EuReDatA recently, the report is to be published and a lecture in this course will be about that exercise. One of the groups had put much effort in the identification of homogeneous populations and achieved a result of a failure rate decreasing with time. Another group identified this as an effect of mixed populations.
The most familiar feature in data analysis in the so called bath tub curve (Figure 5).

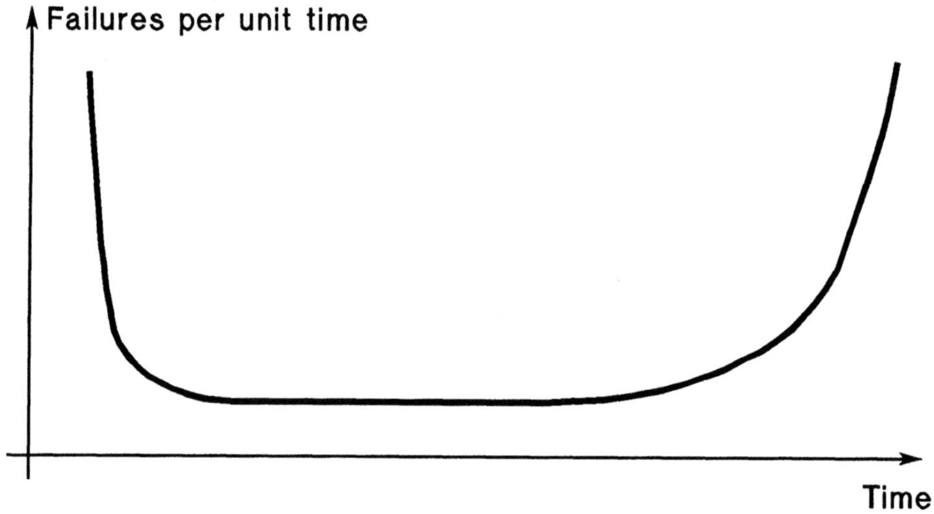

Figure 5: Bath tub curve

The curve represents the behaviour with time of the failure frequency of a homogeneous population of repairable items: a high "failure rate" in the outset due to undetected manufacturing weaknesses, a low and fairly constant failure rate during mid life, and an increasing failure rate due to aging. The striking feature is the section constant in time. I have not yet understood how this can happen. Technical items are deterministically manufactured to work as designed over a period of time. The length of that period may depend on ambient conditions. It is obvious that some may have flaws becoming effective at an early age. It is understandable that aging is affected by several features smoothing the increase of failure frequency at the end of the lifetime. But in between I would expect either a zero failure rate as designed or anything else but a finite constant failure rate.

Table 2: Weibull distribution

Failure probability distribution:

$F(t) = 1 - R(t)$

$R(t)$ = probability function

3-parameter Weibull function:

$R(t) = \exp[-((t-t_o) / (T-t_o))^b]$

t = life time parameter

t_o = "failure free" time

T = characteristic life time: $F(T) = 1 - \exp(-1) = 0.632$

b = shape parameter

Probability density function:

$$F'(t) = \frac{d}{dt} F(t) = a(t)\,[1-F(t)]$$

$$a(t) = b / (T-t_o) \cdot [(t-t_o) / (T-t_o)]^{b-1}$$

Hazard function:

$$\lambda(t) = F'(t) / [1-F(t)] = a(t)$$

Exponential distribution:

$b = 1$, simplification: $t_o = 0$

$F(t) = 1 - \exp(-t/T)$

$F'(t) = 1/T \cdot F(t)$

$\lambda(t) = 1/T = \text{const.}$

One of the frequently used mathematical tools in data analysis is the Weibull distribution (Table 2). Its most simple form is the exponential probability distribution. It is widely assumed, that if a phenomenon - like failure occurrence - is exponentially distributed, this will imply a completely random process behind the phenomenon - as for instance radioactive decay which follows an exponential law and is the manifestation of an entirely random process of the quantum mechanical type. This puzzles me again, because I do not know of such a random process behind failure

occurrence, and I do not see any possibility of quantum phenomena manifestations in failure occurrence. On the contrary, I would expect items which are manufactured and operated in deterministic ways to keep to these ways, even during the small fractions of their life time when they develop failures.
But obviously, they do not. Very surprising, indeed.

3.2 The Need of Data Analysis

As previously outlined, data analysis is the interlink between accumulated information from experience and the application and utilisation of this information. Since data collection is deliberately and purposefully performed in order to utilise the information for the benefit of reliability and safety, the need for establishing the necessary interlink is apparent.

The basic information is quantified and the information required for reliability assessments comprises quantified parameters. In consequence, the interlinking data analysis has to transform basic data into reliability parameters. The transformation means condensation; for instance:

- Homogeneous groups of items are identified within component classes.

- For these samples the time behaviour of failure occurrence and repair times is considered in such a way, that the hazard and repair rates are derived and quantified as functions of time.

- According to the specific statistical methods used for the derivation of the rates, the results are averaged quantities - mean or median values, for instance. The individual quantities from which the averaged ones are derived scatter around the averages.

- According to the particular way of scattering, a distribution function can be allocated which quantitatively describes how the individual data scatter, what the appropriate averaged quantity is and what uncertainties are to be allocated to the average.

- It sometimes happens that the derivation of rates does not work, because the data does not fit any of the distributions used or show otherwise strange behaviour. Before one tries more exotic or strange distributions, it is advisable to check the basic data again for possible flaws like inhomogenities or even non-random "contaminations". The best way to do the check is pattern search.

- As mentioned before, the 3-parameter Weibull function (Table 2) is a highly regarded and frequently used tool for the condensation of basic data into rates. The reason for this is that it is a very flexible and versatile tool as can be seen in the figures 6, 7 and 8 which are all based upon the same parameter sets and show the failure probability function F(t), the probability density function F′(t) and the hazard function λ(t) - respectively. The parameter t_o represents a failure free period of time, which is of increasing importance for product warranties. The case of exponential behaviour is included and characterised by parameter b = 1.0, most clearly shown in figure 8 as the constant rate. The probability density function F′(t) in figure 7 shows rather clearly the variety of distributions which may be represented by the Weibull approach.

It should be emphasised here that there is a serious disadvantage in the Weibull approach - as in most statistical approaches in reliability engineering. If it works, it may give you the feeling you understand what is going on in reality. Don't adopt this feeling as a conviction, it is a feeling without any confirmation at all. The statement requires some comment:

- If the Weibull or any other approach does not work, one usually has no idea why and starts a more or less incoherent search for a solution. Now, if it works, one is at the same level of knowledge except that one feels no need to search for a solution.

- The only parameter in Weibull directly referring to reality is the failure free period t_o, which in many cases is rather near zero. All other parameters do not carry any perceivable information about reality, they are used for fitting and deriving reliability parameters describing the behaviour of a sample, nothing else.

- The only exception is the case b = 1, i.e. exponential behaviour. According to usual interpretation, exponential behaviour represents pure randomness, i.e. information, knowledge, understanding are zero. This is the most appropriate description of reality and should be kept in mind if one develops some strange feelings of understanding. Despite this disastrous situation, you can work with it.

I expect these comments will raise some argument.

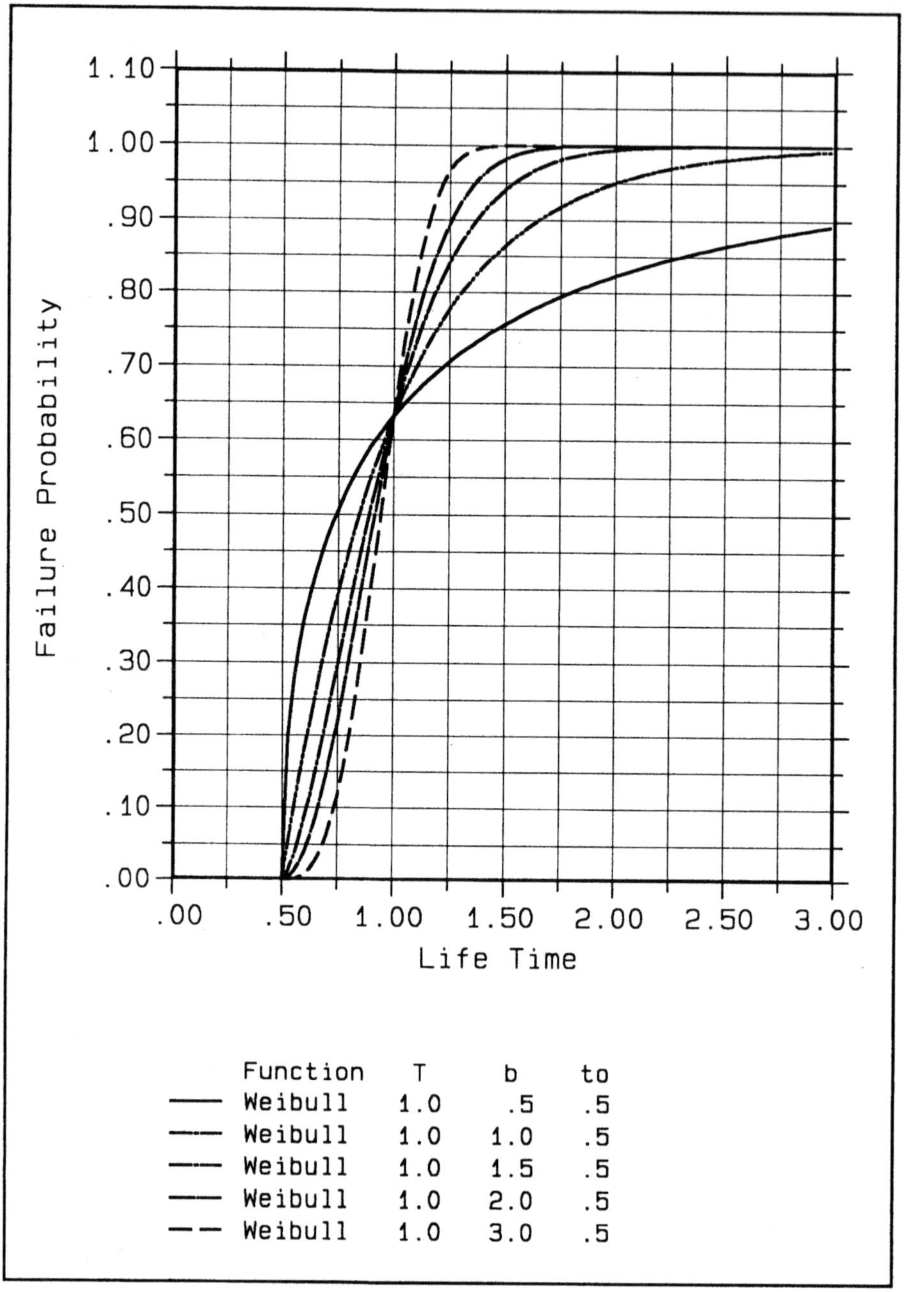

Figure 6: Failure probability parameter variation F (t)

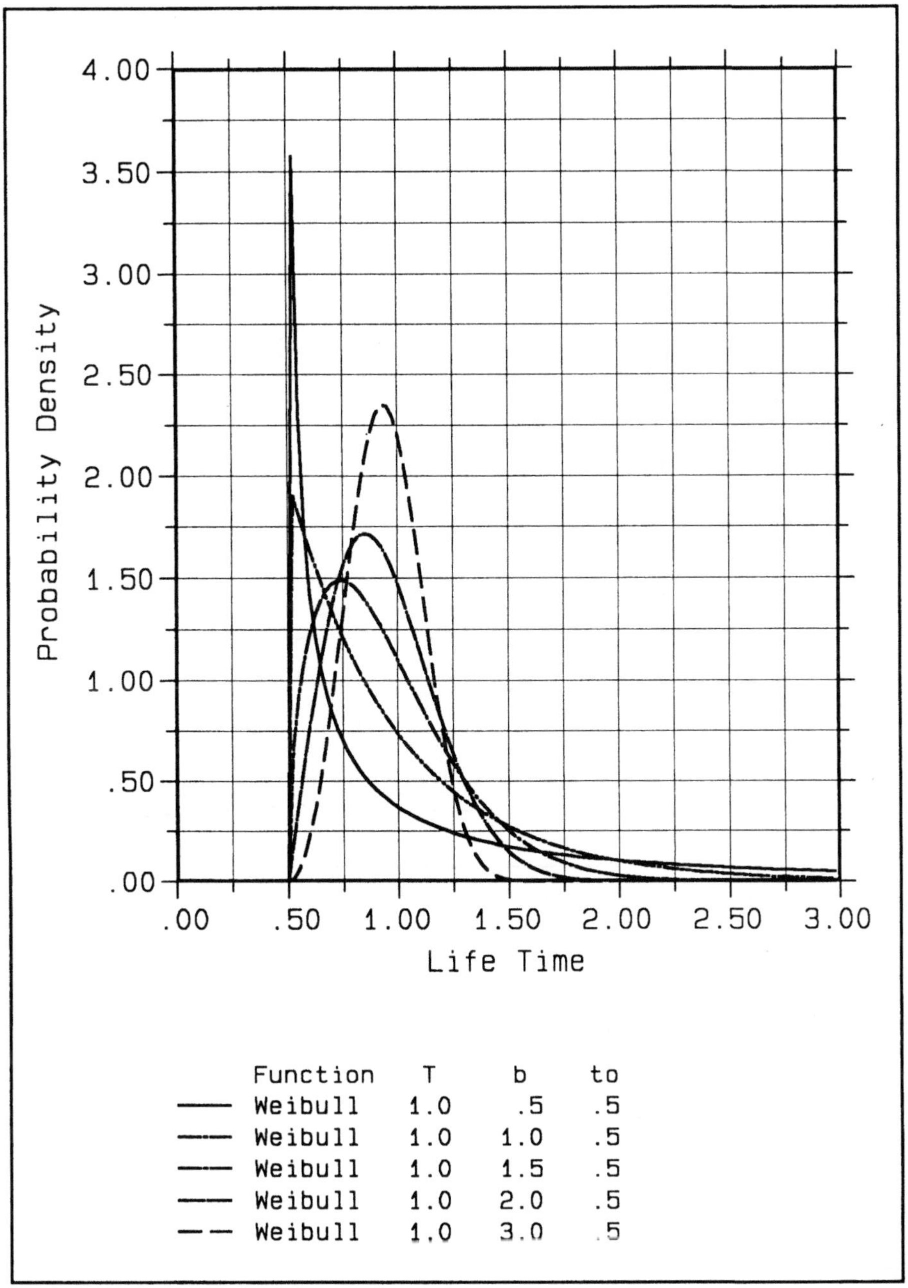

Figure 7: Probability density F′ (t)

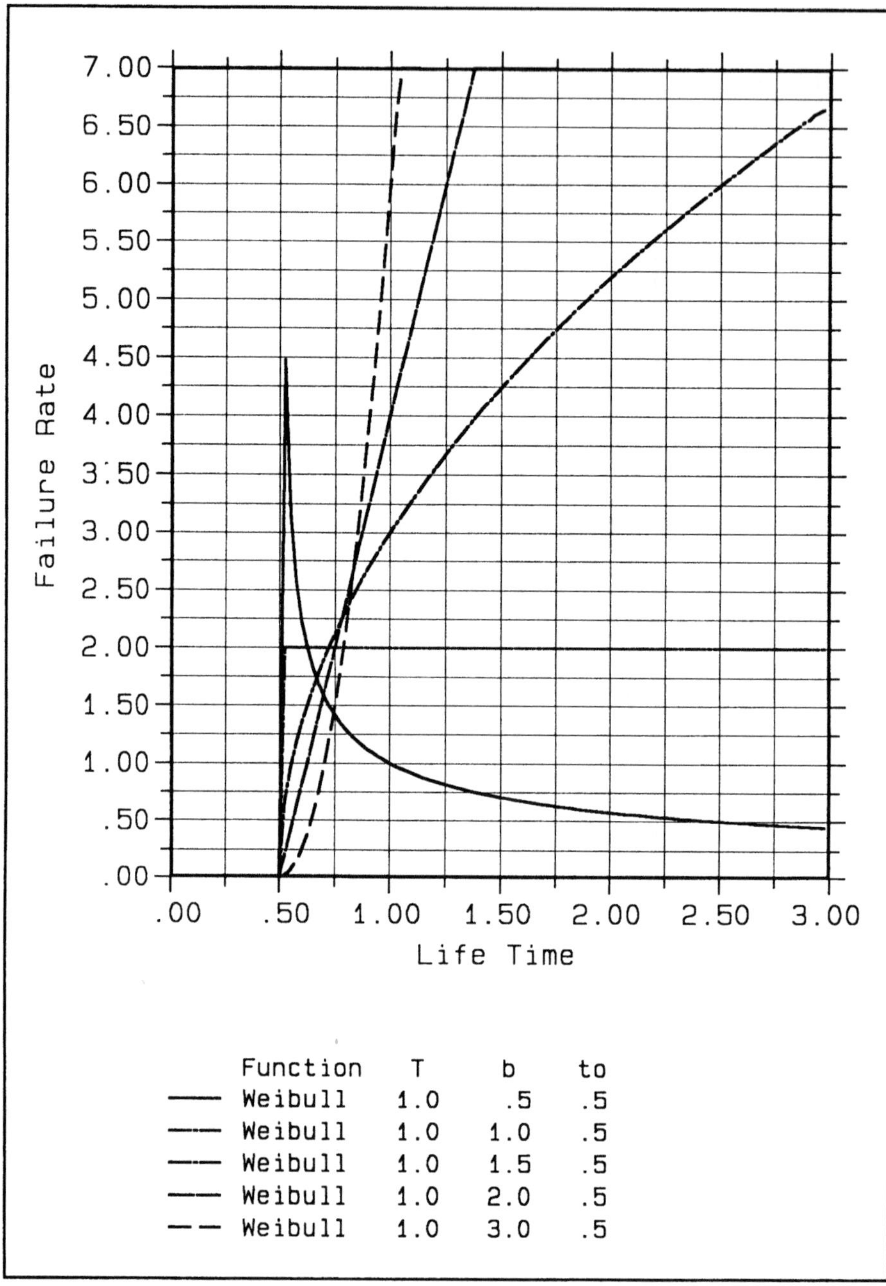

Figure 8: Failure or hazard rate presentation λ (t)

3.3 The Use of Data Analysis

The use, i.e. the performance of data analysis is determined by its objective; i.e. supplying quantified information meeting the demands of reliability engineering. The supply must comprise more than a mere set of parameters. It must include:

- the boundary conditions under which the information was obtained,
- the probable limits of applicability which may be set by the environment,
- the mode of operation etc. of the original items which form the data source,
- the uncertainties of the parameters with the confidence levels. An appropriate documentation of this information should also contain the methods with which it was derived.

There are, however, some limitations, with which one has to unfortunately live. It has not yet become common practice to use time dependent failure rates in reliability engineering. The reason is that it can be rather difficult and computer time consuming to evaluate fault trees if the parameters are time dependent. It is still more difficult to process the uncertainties through a fault tree, when both the parameters and their uncertainties are time dependent. Therefore, simplifications are often necessary, which sometimes mean introducing the assumption of constant failure rates and uncertainties.

A second limitation may come from the sheer lack of sufficiently abundant basic data. The smaller the set of basic data, the higher the uncertainties and the more probable a good fit with an exponential law. Thus, it ends again at constant failure rates, with which the reliability engineer may be rather pleased.

He/she should not be too pleased, because he/she is urgently required to justify his/her simplifications or the use of a scarce data set. This means, that one is supposed to give an account of how the results might probably alter with more realistical conditions and what more realistical conditions are, which might not always be easy.

Figure 9 shows a good example of a 3-parameter Weibull fit of the front brake lining data from a sufficiently ample and homogeneous set of cars (this can be concluded from the uncertainties) over a life time of 100.000 km. As the derived probability function exceeds the exponential function, the failure rate is increasing with time. The failure free period of the linings is near to 9.000 km.

Figure 10 shows a failed attempt of fitting car dashboard instrument failures with a Weibull approach, although the data are again sufficiently ample and the instruments of a really homogeneous population. The error bars lay within the data dots.

I refer to this example because it stresses the importance of data pattern search, in order to reveal inconsistencies or contaminations. Pattern search means the plotting of failure frequencies against various parameters like life time in km, in months, against calendar time or geographical regions.

In this particular example it was calendar time which revealed an obviously non-stochastic pattern (Figure 11): a peak occurring at a specific time in the year for all monthly batches of production independent of the month of production. Each horizontal line represents the monthly failures of a month's production of a specific type of instruments. There is a second peak occurring about two to tree months after the time of commissioning. When we investigated the matter further, the calendar fixed peak turned out to occur at the summer holiday time, when everybody has all parts of their cars fixed at the workshop, even those parts which one does not care about at other times. The peak at the earlier time obviously represents a warranty effect at first report to the workshop.

This example is a case of man made data contamination, which does not apply to brake linings. Vital failures are reported immediately after detection or even suspicion.

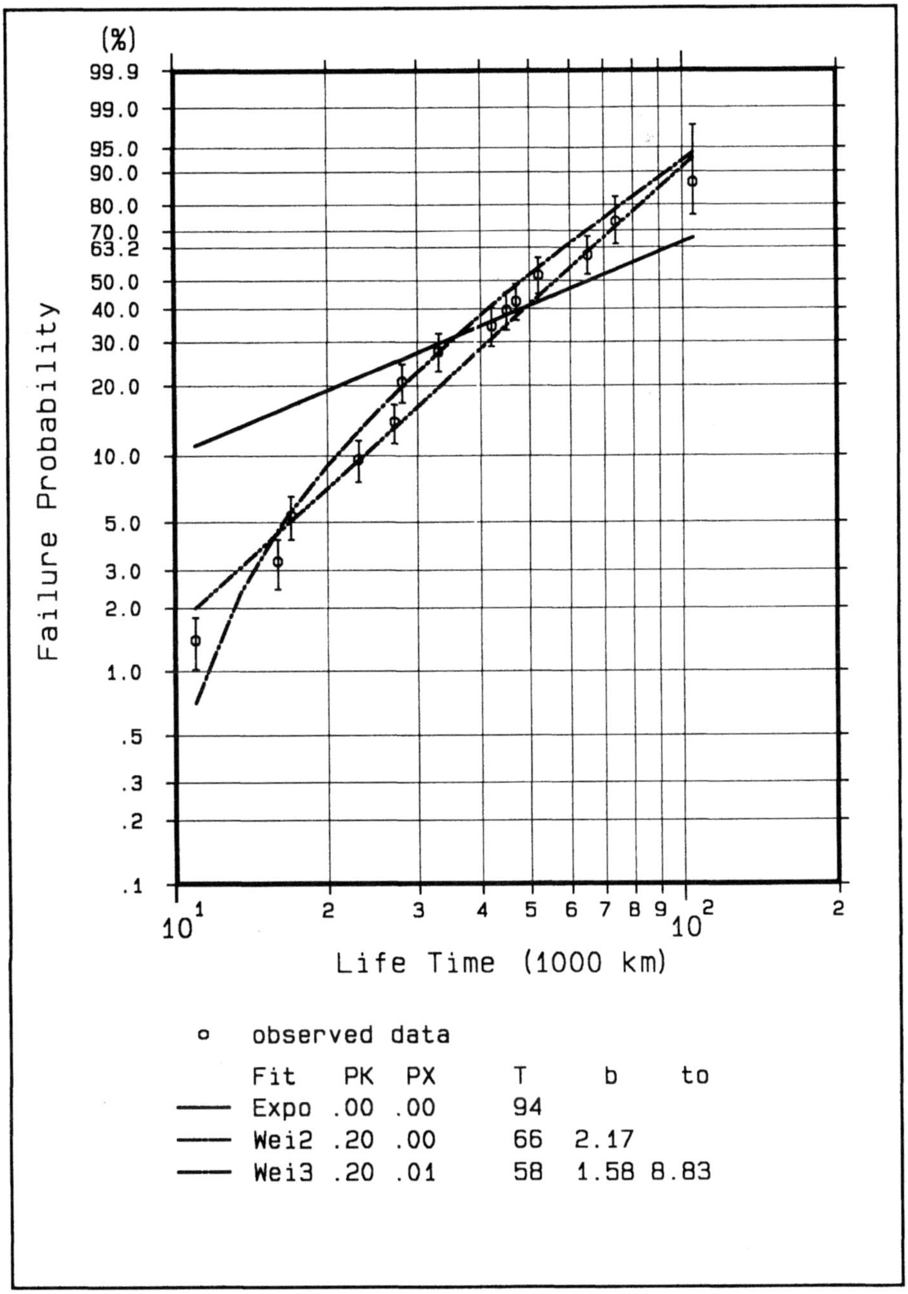

Figure 9: Fit of front brake linings failure data - indicating the values of the fit parameters and of the fit test results, ref. /8/
(PK: Kolmogorow-Smirnov, PX: Chi sqare)

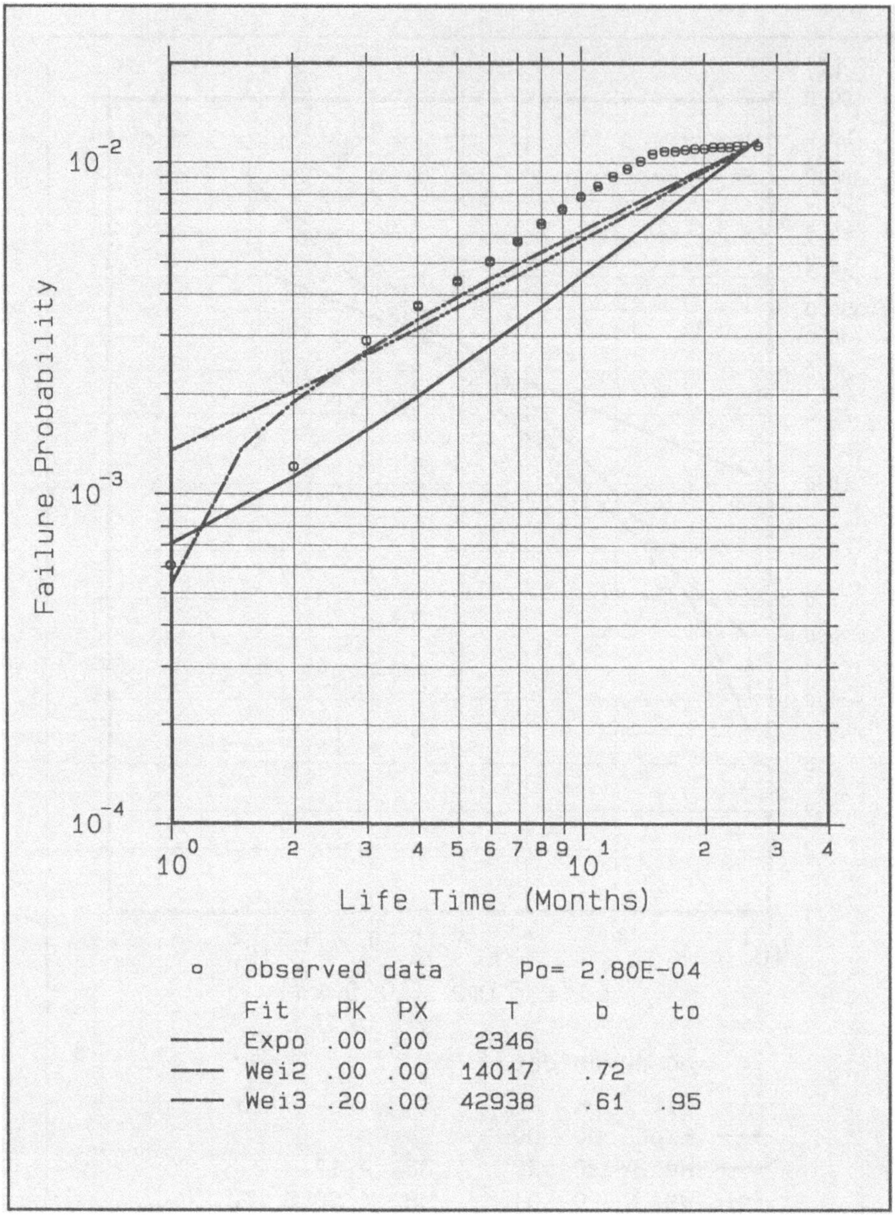

Figure 10: Unsuccessful attempt of a failure data fit of instruments failures, ref. /8/

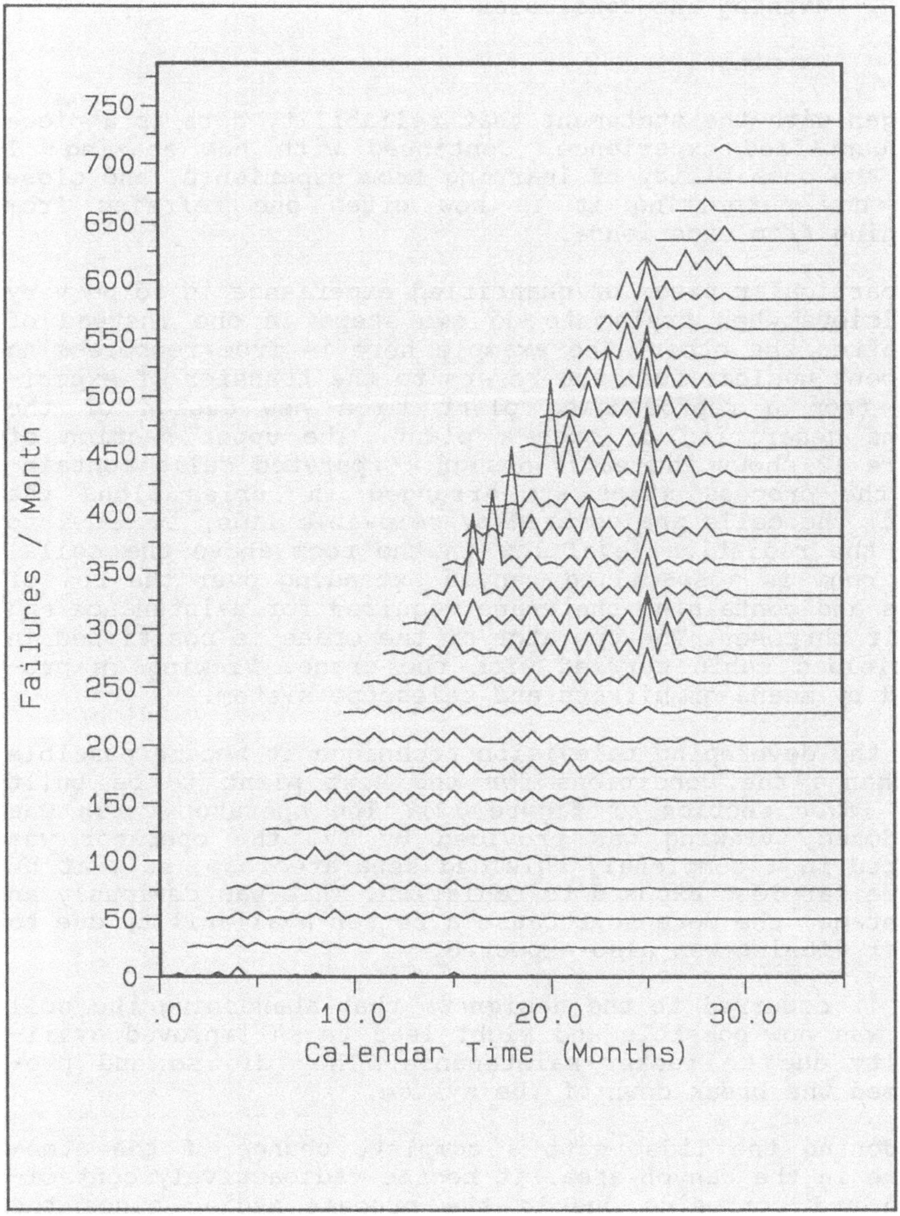

Figure 11: Failure data pattern revealed, ref. /8/

4. Warning and Conclusion

4.1 Warning

I began with the statement that reliability data is a piece of quantified experience, continued with how amazing I find the possibility of learning from experience, and close with how surprising it is how often one refrains from learning from experience.

One particular piece of quantified experience is to be very suspicious when trying to do two steps in one instead of one after the other. The example here is from reprocessing of spent nuclear fuel and refers to the transfer of experience from a reprocessing plant to a new design of the second generation of such a plant. The upper section of figure 12 shows the early design: Separated cells containing the process steps are arranged in series along the plant; the cells are covered by removable lids, in order to keep the radiation level low in the room above the cells. This room is a so-called canyon extending over the row of cells and containing the crane required for maintenance and repair purposes. The operator of the crane is positioned in a shielded cabin carried with the crane. Viewing is provided by means of mirrors and telescope system.

With the developing television technique it became possible to change the conditions for the next plant to be built (see lower section of figure 12). The operator cabin was abandoned, viewing was provided by TV, the operator was located in a completely shielded separate room, so that he became far less exposed to radiation. This was obviously an advantage, the more so because a better availability due to better viewing was also expected.

Then it occurred to the designers that abandoning the cell lids was now possible and might lead to an improved availability due to faster maintenance. They did so and programmed the break down of the system.

Abandoning the lids meant a complete change of the atmosphere in the canyon area: it became radioactively contaminated and corrosive due to the process media, since the ventilation system had to be changed. In the former design the air flow was from the canyon into the cells. This was not possible any more without lids.

In consequence, the crane was now to operate in a hostile environment which it was not fit for. Corrosion occurred at cables, end switches, motor and gearing. It crashed occasionally against the front wall. Repair of the crane - the

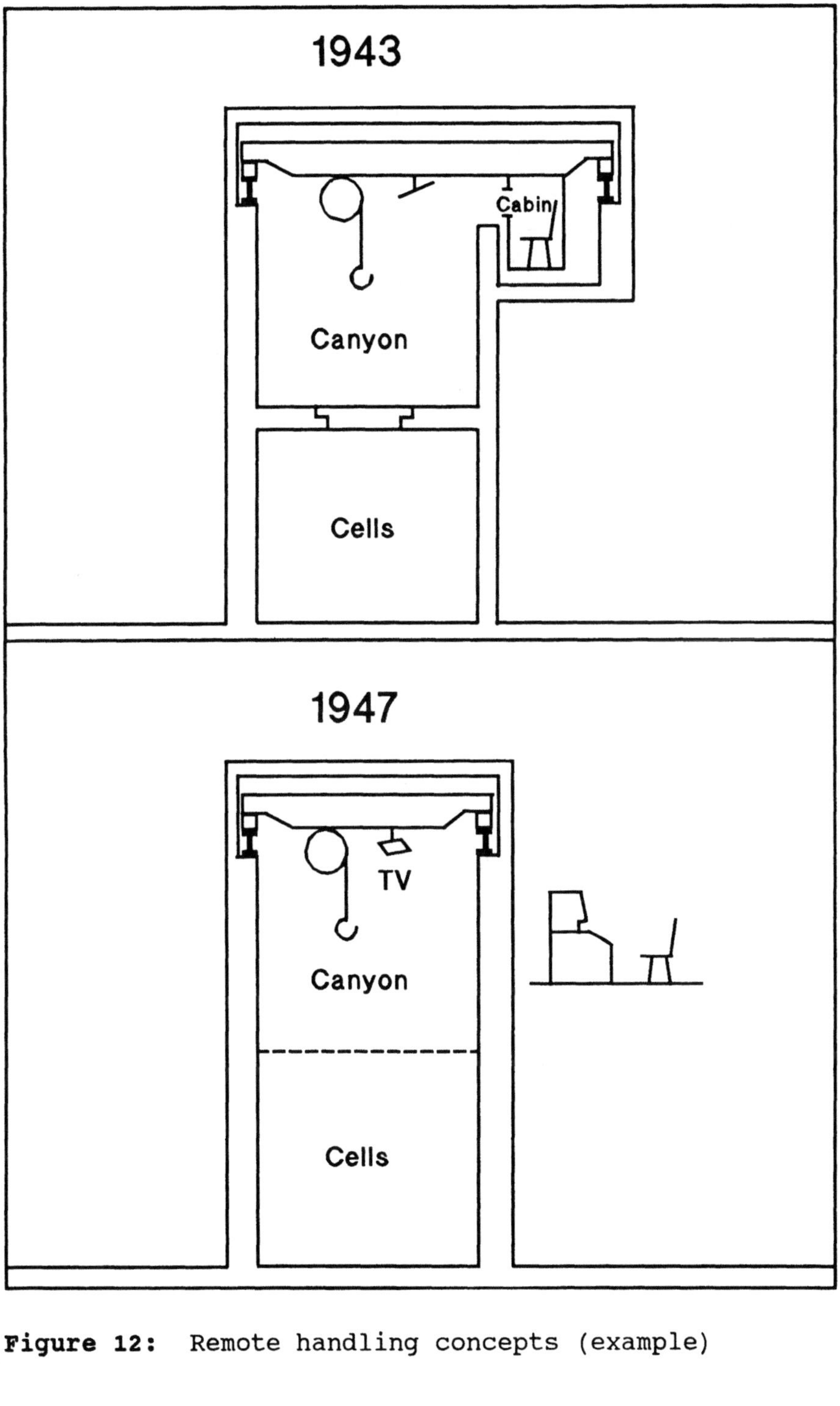

Figure 12: Remote handling concepts (example)

only maintenance tool - became more and more frequent and more and more difficult, because it became heavily contaminated. Repair times increased considerably. Decreasing repair rate with increasing failure rate is disastrous for the availability of an item. If this item is indispensable for the operability of the system, the latter is certainly endangered. And so it happened: the same effect as with the bridge near Kufstein. In the example, it became worse: the more frequent and more extended repair of a more contaminated crane raised the dose to the staff considerably. The initially intended improvements of availability and safety had disastrously failed.

4.2 Conclusion

In the beginning I promised to outline some questions and problems. Although not all of them are very easily identified, I am sure you have found them.

I have deliberately used some opacity or ambiguity in my formulations as those questions and problems tend to change character when viewed from different angles. Terms formulated too definitely may seriously bias the viewing. It was however intended to promote discussion and communication, which require free viewing. I hope I have met the expectations.

5. References

/1/ Stevens, B., Editor (NCSR, U.K.)
Guide to Reliability Data Collection and Management, EuReDatA Project Report no. 3,
CEC-JRC Doc. no. S.P./I.05.E3.86.20
Ispra, 1986

/2/ OREDA Participants, Editors
OFFSHORE Reliability Data Handbook,
VERITEC, Høvik, N, 1984

/3/ Military Standardization Handbook; Reliability Prediction of electronic Equipment, MIL-HDBK-217 C,
U.S. Department of Defence, 1979

/4/ Balfanz, H.,
Ausfallratensammlung
Report IRS-W-8, 1973
GRS, Köln, D, 1973

/5/ Luisi, T., Coordinator (CEC-JRC Ispra, I)
Reference Classification Concerning Components' Reliability,
EuReDatA Project Report no. 1
CEC-JRC Doc. no. S.A./I.05.01.83.02
Ispra, 1983

/6/ Garnier, N., Coordinator (CNET, F)
Proposal of a Minimum Set of Parameters in Order to Exchange Reliability Data on Electronic Components,
EuReData Project Report no. 2,
CEC-JRC Doc. no. S.P./I.05.E3.85.25
Ispra, 1985

/7/ Gavelli, G., Smith, A.L., Editors
Materials Reliability
EuReDatA Project Report no. 4
CEC-JRC Doc. no. S.P./I.05.E3.85.38
Ispra, 1985

/8/ Leicht, R., Oehmke, R., Wingender, H.J.
Collection and Analysis of Car Instrument Failure and Survival Data,
Proceedings of the 7th International Conference on Reliability and Maintainability,
Brest, F, 1990

RELIABILITY - AVAILABILITY - MAINTAINABILITY - DEFINITIONS

OBJECTIVES OF DATA COLLECTION AND ANALYSIS

A. LANNOY
EDF - Groupe Retour d'Expérience
Département REME
25, allée Privée, Carrefour Pleyel
93206 SAINT-DENIS CEDEX 1

Summary
The aim of this paper is to specify the terminology relating to reliability, availability and maintainability. It shows the interest of creating feedback of experience files, in particular for applications relating to the safety and maintenance of installations.

1. Maintenance and Medicine

Before examining the concepts of maintenance, reliability and availability, it is relevant to draw a comparison between human health and the lifespan of a piece of equipment, as shown in figure 1.

Analogy

HUMAN HEALTH			HEALTH OF THE MACHINE
	Birth	Commissioning	
Knowledge of humans			Technological information
Knowledge of illness	Long-life	Durability	Knowledge of failure modes
Health record			History
Medical file			File on the machine
Diagnosis, examination			Diagnosis, expertise
Visit	Good health	Reliability	Inspection
Knowledge of treatments			Knowledge of curative actions
Curative treatment			Overhaul, repairs
Operation			Renovation, modernisation, exchange
	Death	Scrapping	
MEDICINE			INDUSTRIAL MAINTENANCE

Figure 1 - Analogy between human health and the lifespan of a piece of equipment

J. Flamm and T. Luisi (eds.), Reliability Data Collection and Analysis, 45–59.

The analogy is obvious. Many similitudes exist as regards:

- monitoring (optical or acoustic monitoring systems, endoscopes, etc.),
- inspection and checks (X-rays, ultrasonic tests, etc.),
- diagnosis and assistance possibly provided by artificial intelligence,
- data banks on history (feedback of experience banks, health record) and their analysis (consultation, statistical analysis, multivariate analysis, etc.).

The life and maintenance of a piece of equipment begin at design stage, at which time maintainability (ability to be maintained), reliability and availability (ability to be operational) and predicted life time are determined.

Follow-up of a piece of equipment provides a clearer view of its behaviour, its weaknesses and the nature of degradations, etc., as well as any information which may lead to improvements in equipment (changes in design, maintenance for improvement), and an optimisation of the maintenance strategy on the general basis:

- either of a probabilistic safety criterion,
- or of an economic criterion aiming at minimising the ratio:

$$\frac{\text{maintenance costs} + \text{forced shutdowns}}{\text{service rendered}}$$

2. The concept of maintenance

2.1. Definition of maintenance

Maintenance can be defined as a set of actions implemented to maintain or restore a link to a specified state or in a condition in which it can provide a given service.

In this definition, we have the ideas:

- of prevention (keeping a system or a unit of equipment in working condition),
- of correction (concept of restoring),
- of service (specified state or given service).

Different types of maintenance can be defined, as shown in figure 2.

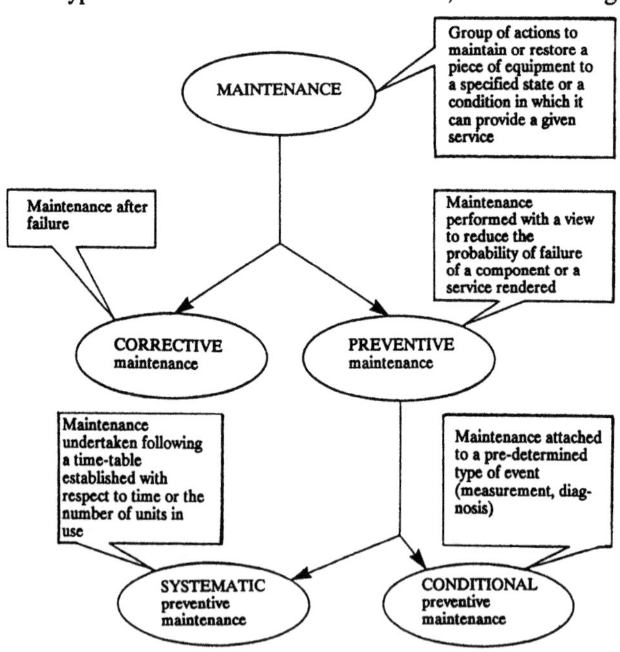

Figure 2 - The different types of maintenance

Corrective maintenance is performed after a failure. It can be palliative (case of overhaul) or curative (case of repair). Maintenance can be *preventive*, i.e. conducted with a view to reducing the probability of failure or the deterioration of a service rendered. It can be systematic following a time-table, in which case, it is assumed that the behaviour of equipment is known over time. It can be *conditional* (predictive), i.e., subordinated to a pre-determined type of event: possible ongoing follow-up of a unit of equipment, existence of a progressive and measurable deterioration, correlation between a measurable parameter and the state of equipment.

2.2. Failure

The definitions of maintenance lead to a definition of failure, which corresponds to the alteration of a unit of equipment or its stopping to fulfill a required function (synonyms sometimes used depending on professional sectors: damage, wreckage, anomaly, breakdown, incident, fault).

Failure can be:

a) as regards the degree of failure:
   - partial: alteration of operating conditions,
   - total: operation stops, or
   - critical: total failure which has repercussions on the safety of the installation;

b) as regards the appearance of the failure:
   - catalectic: sudden and total,
   - through deterioration: progressive and partial;

c) as regards the trend of the failure rate:
   - random: constant failure rate $\lambda(t)$,
   - infant mortality: decreasing failure rate $\lambda(t)$,
   - wear out: increasing failure rate $\lambda(t)$.

The failure rate $\lambda(t)$ (which is a reliability estimator) represents a proportion of faulty systems (number of failures/duration of use).

Figure 3 portrays the trend in the failure rate over time (age) for electronic and mechanical equipment.

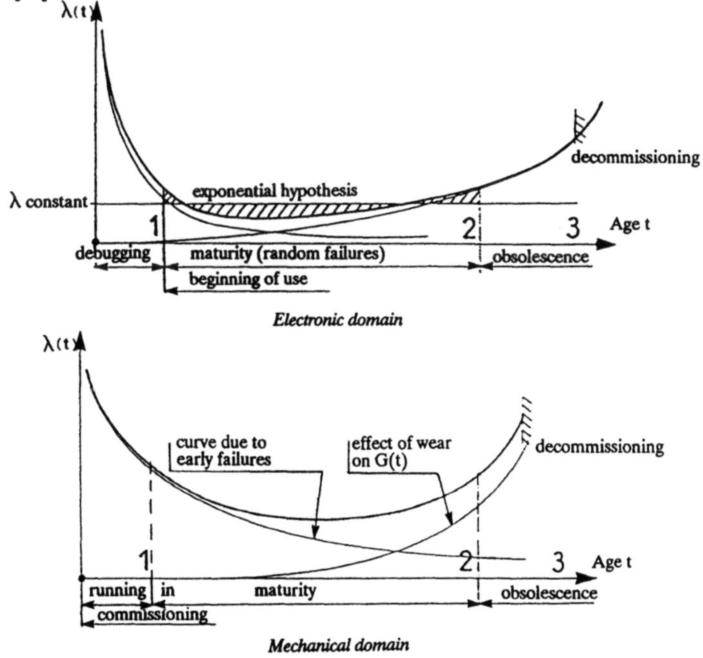

Figure 3 - The bath curve of reliability experts

The deterioration process is frequently as follows:

initiation ----------> propagation ----------> loss of the function.

This process is initiated by a *cause of failure*, the physical reason for which one (or several) internal component(s) is (are) deteriorated, thus causing failure of the component. In the case of mechanical failure in service, this cause can be due to collision, overload, thermal or vibrational fatigue, creep, wear, abrasion, erosion, corrosion, etc. In the case of electrical failures, it can be due to rupture of the electric link, breakdown, sticking, wear of contacts, etc.

The *failure mode* is the occurrence of an abnormal physical phenomenon through which the loss or the risk of loss of the function of a given unit of equipment is observed.

2.3. The concepts of reliability, availability and maintainability

They are illustrated in figure 4.

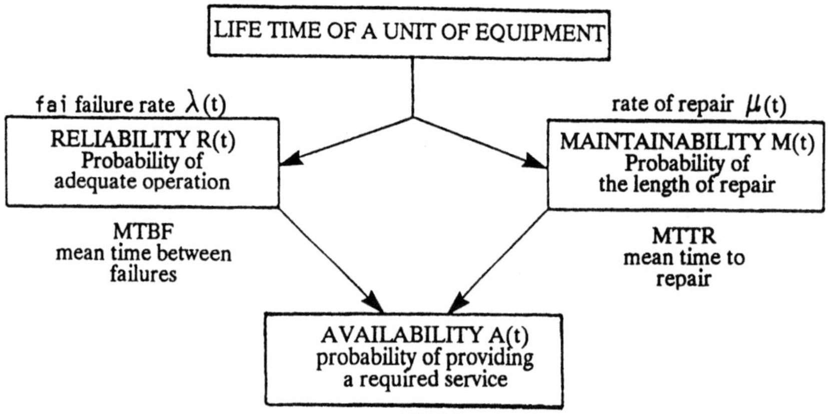

$$A = \frac{MTBF}{MTBF + MTTR}$$

Figure 4 - The concepts of reliability, maintainability and availability

Reliability is the characteristic of a device expressed by the probability that this device will perform a required function in the given conditions of use over a determined length of time. It is designated by R(t) which is therefore the probability of adequate operation.

A distinction should be drawn between quality and reliability: *quality* is the conformity of a unit of equipment with its specifications, whereas reliability is an extension of quality over time: the unit of equipment should remain in conformity with its specifications throughout its life time.

MTBF (mean time between failures) is a characteristic of reliability: it corresponds to the mean of adequate operation between consecutive failures (or the mathematical expectation of the mean variable: date of appearance of the failure).

Similarly, *maintainability* is the probability that the device will be restored to a given operating state in a given time after failure. It is characterised by a *MTTR* (mean time to repair).

Availability is the probability that the device will be in operating condition. The device is therefore neither in failure or maintenance mode. Availability is therefore a function of reliability and maintainability. Increasing the degree of availability consists of reducing the number of failures (action on reliability) and reducing repair time (action on maintainability).

2.4. Knowledge of equipment

All of the failures mentioned earlier reveal the need for accurate knowledge of a unit of equipment throughout its life cycle.

Figure 5 illustrates all the concepts listed above.

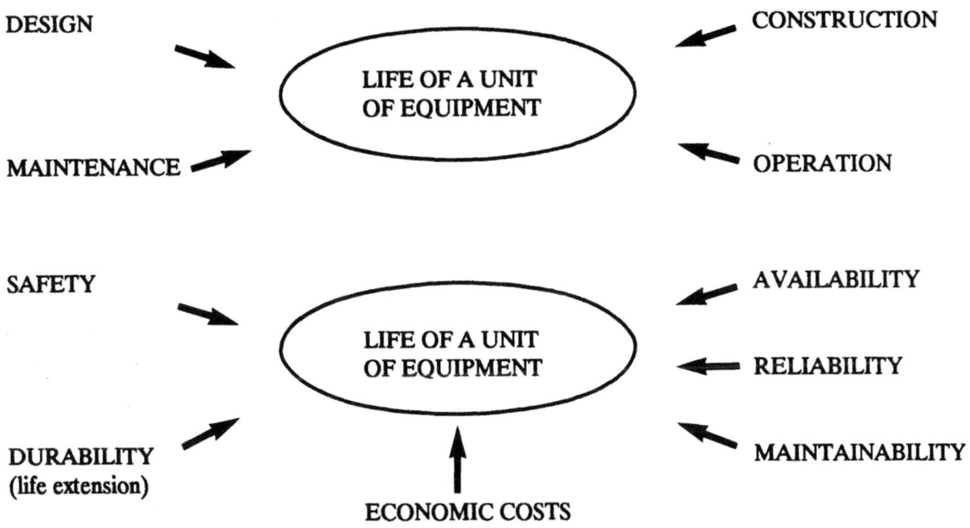

Figure 5 - Knowledge of the life of a unit of equipment

In general, the operator of a unit of equipment makes an attempt to find correlations between the state of this equipment and a measurable parameter:

- physical parameters (temperatures, pressures, etc.),
- vibration or acoustic emission levels, ultrasonics,
- analysis of lubricants,
- etc.

In addition, all failures, events, preventive or corrective repairs of equipment as from industrial commissioning must be described in chronological order.

Feedback of experience files play this basic role.

3. Feedback of experience banks

The paragraphs below are based on Electricité de France's experience.

Different data banks form the feedback of experience of installation and equipment operation and are used to memorise experience gained. They consist of:

a) The data banks of tests relating to local physical phenomena having occurred on sensitive parts of the installation:
   - test data comprising physical magnitudes (pressure, temperature, deformations, etc.),
   - operating parameters recorded on line during operation (monitoring data banks, transients),

- data derived from inspection and checks, comprising measurements of wear or of the state of the material to be able to follow its degree of damaging or wear (eddy current, ultrasonic, etc., checks).

b) The operating results of installations, listed in the data banks relating to:
  - availability (events file),
  - reliability (failure file as in the Reliability Data Collection System (SRDF)),
  - statistics (operating statistics file),
  - maintenance (maintenance operations history file).

c) Constructive data required for the assessment of factors acting in the deterioration process.

Figure 6 illustrates these principles. Considering the reliability-related character of this text, the follow-up of the paper examines only the banks referring to reliability and availability.

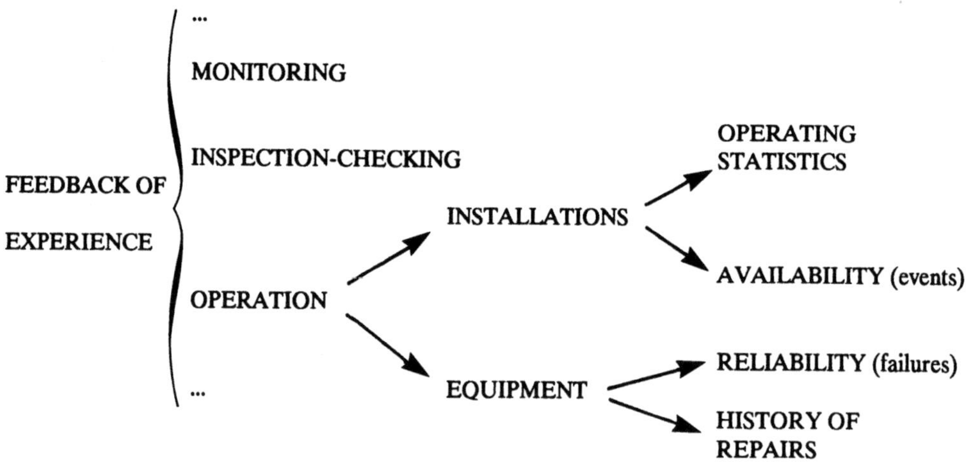

Figure 6 - Feedback of experience data banks

4. Objectives

Needless to say that the creation of these files is expensive. As a result, a high return is expected.

In reality, the importance of a bank (and its interest) depends on the critical character of the installation or equipment.

The creation of these files depends on the reliability, availability and maintainability objectives attributed to them.

The prime target is quite frequently safety-related and approval of installations to be able to identify on the one hand, the critical failures observed on safeguard equipment and, on the other, the serious forewarning initiating events, to provide information for the probabilistic safety studies.

It emerges then that although these files are quite frequently built for safety purposes, they can be put to various other uses:

- reliability: determination of reliability laws, failure rates, optimum repair times,
- availability: appraisal of availability coefficients of the installation, equipment,
- methods: search for and selection of sensitive (or critical) components,
- stock management,

- maintenance policy: optimisation of the most suitable maintenance policy for the component, which is only possible when the history (failures and repairs) of the component is available,
- decision-making assistance: cost-benefit analyses,
- equipment design assistance: disclosing of the critical components (and subcomponents) and damaging mode, application to design, modifications and improvements of components (improvement maintenance).

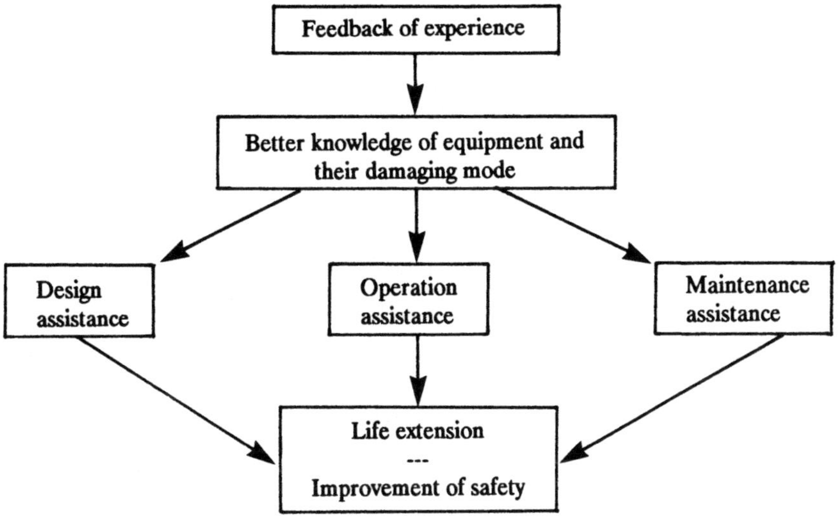

Figure 7 - Contribution of the feedback of experience

5. Creation of data banks

5.1. Availability files

This creation depends on the use that is to be made of them.

As an example of an availability file, figure 8 shows an "event form" taken from the "events file" of French PWR plants. This file is basically used for managing the feedback of experience. All events relating to the operation of units are recorded in this file, especially:

- all turbine trips due to an incident which has occurred either inside or outside the plant,
- all equipment failures observed in service or when shut down,
- all events deemed to be significant from a safety point of view, following criteria selected by the Safety Authorities.

The information is collected in the power station. It is then memorised. Each file gives a factual description of each event. It should be observed that many items are coded. In addition, the free summary, often a mine of information, is important since it lists the following type of information: consequence - circumstances with chronology - causes, repairs, involving action - reference (this information is generally specified in this syntaxic order).

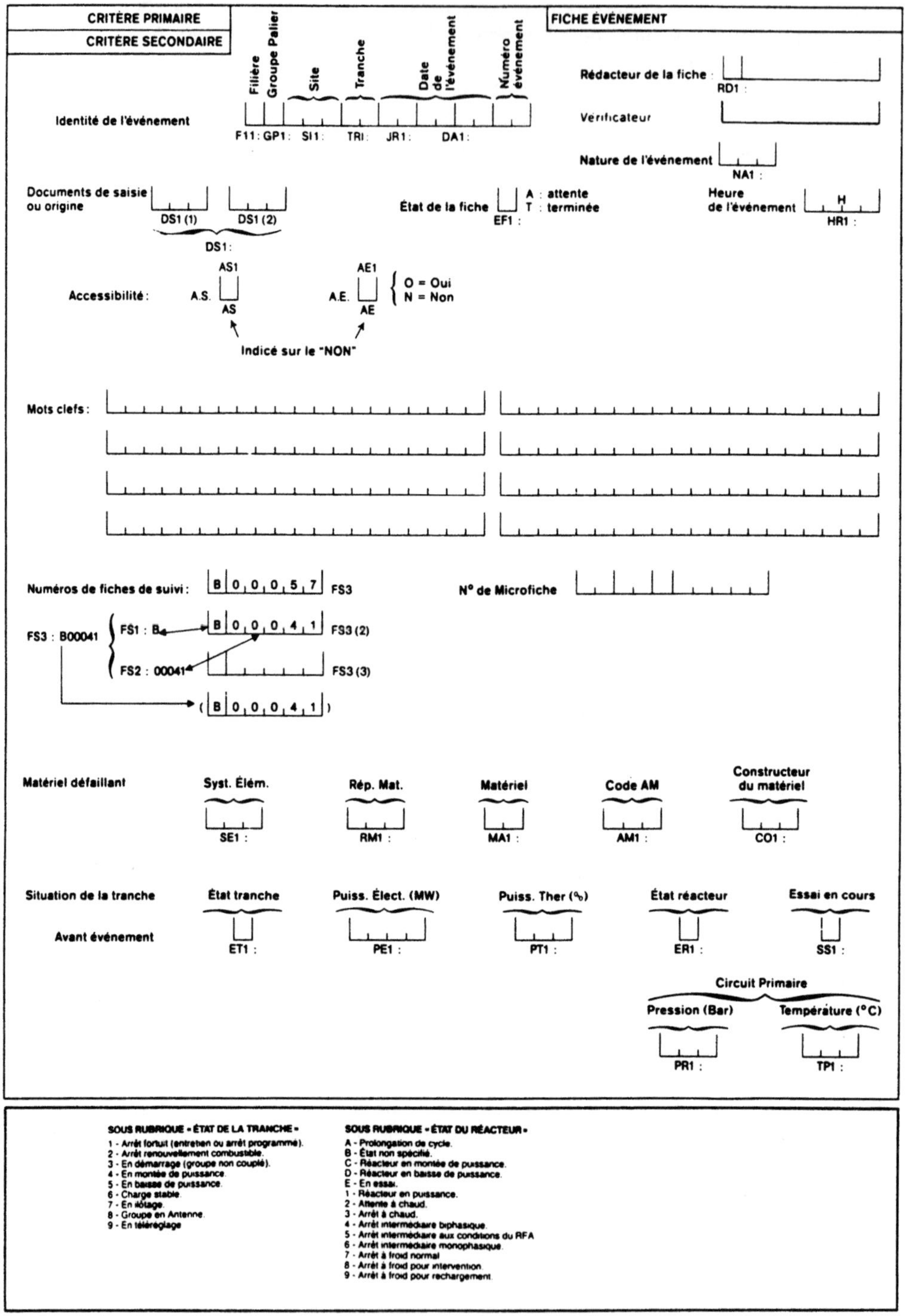

Figure 8 - The events form - French PWR units

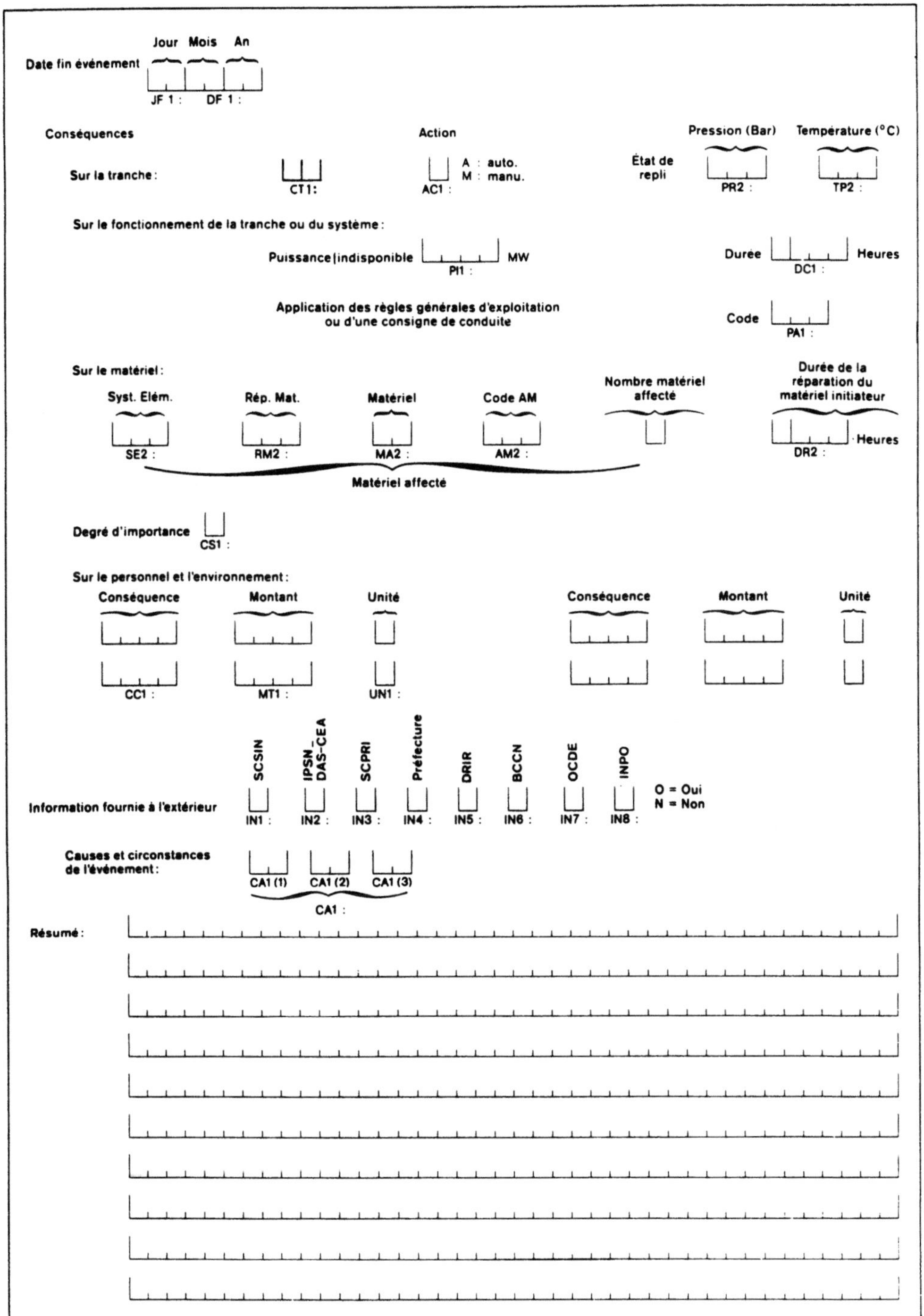

Date fin événement — Jour / Mois / An — JF 1 : DF 1 :

Conséquences

Sur la tranche : CT1:

Action — AC1 : A : auto. M : manu.

État de repli — Pression (Bar) PR2 : — Température (°C) TP2 :

Sur le fonctionnement de la tranche ou du système :

Puissance indisponible PI1 : MW

Durée DC1 : Heures

Application des règles générales d'exploitation ou d'une consigne de conduite

Code PA1 :

Sur le matériel :

Syst. Elém. SE2 : — Rép. Mat. RM2 : — Matériel MA2 : — Code AM AM2 :

Matériel affecté

Nombre matériel affecté

Durée de la réparation du matériel initiateur DR2 : Heures

Degré d'importance CS1 :

Sur le personnel et l'environnement :

Conséquence CC1 : — Montant MT1 : — Unité UN1 :

Conséquence — Montant — Unité

Information fournie à l'extérieur

SCSIN IN1 : — IPSN_DAS-CEA IN2 : — SCPRI IN3 : — Préfecture IN4 : — DRIR IN5 : — BCCN IN6 : — OCDE IN7 : — INPO IN8 :

O = Oui
N = Non

Causes et circonstances de l'événement : CA1 (1) CA1 (2) CA1 (3) — CA1 :

Résumé :

Figure 8 (continued) - The events form - French PWR units

☐ AFFICHAGE ☐ ÉMISSION

☐ MODIFICATION ☐ SUPPRESSION

350 CLÉ – TRANCHE – ANNÉE – NUMÉRO (T AA NN NN) T A A N N N N

352 REPÉRAGE FONCTIONNEL

354 NUMÉRO DT

162 NUMÉRO CONSIGNATION

100 RÉFÉRENCE DOCUMENTATION

351 DATE DÉCOUVERTE ANOMALIE J J M M A A

490 DATE DÉBUT DÉFAILLANCE J J M M A A

371 ÉTAT DU MATÉRIEL

☐ A ARRÊT

☐ F FONCTIONNEMENT

☐ S SOLLICITATION

SITUATION DU MATÉRIEL

☐ E ENTRETIEN / REQUAL

☐ N SERVICE NORMAL

☐ T ESSAIS PÉRIODIQUES

365 DESCRIPTION SOMMAIRE DE LA DÉFAILLANCE

367 Nb HEURES CUMULÉES

101 DOSE ABSORBÉE (H. mREM)

ECRAN 2

377 SITUATION RÉACTEUR LE JOUR DE LA DÉCOUVERTE DE L'ANOMALIE

☐ 01 EN PUISSANCE > 2 %

☐ 03 ATTENTE / ARRÊT A CHAUD

☐ 04 ARRÊT POUR RECHARGEMENT

☐ 06 ARRÊT A FROID, CUVE FERMÉE

☐ 07 ARRÊT INT. NON CONNECT. AU RRA

☐ 09 ARRÊT INT. SUR RRA

☐ 12 ARRÊT POUR INT. CUVE OUVERTE

379 CONSÉQUENCE DE LA DÉFAILLANCE LE JOUR DE LA DÉCOUVERTE DE L'ANOMALIE

☐ 01 DÉCLENCHEMENT G.T.A.

☐ 02 CHUTE DE BARRES

☐ 04 ILOTAGE

☐ 06 PASSAGE ATTENTE / ARRÊT A CHAUD

☐ 07 RETARD AU COUPLAGE

☐ 08 AUCUNE CONSÉQUENCE SUR PUISSANCE

☐ 11 PERTE VOIE

☐ 12 PERTE TOTALE SYS. OU VOIE

☐ 30 RÉDUCTION DE CHARGE

☐ 40 APPLICATION 10

☐ 41 PASSAGE A L'ÉTAT REPLI (10)

NOTA

403 DEGRÉ DE DÉFAILLANCE ☐ C COMPLÈTE ☐ D PARTIEL

404 APPARITION DÉFAILLANCE ☐ A SOUDAINE ☐ E PROGRESSIVE

339 MODE DE DÉFAILLANCE

340 COMPOSANT INTERNE AFFECTÉ

341 CAUSE DE LA DÉFAILLANCE

364 DURÉE RÉPARATION (H)

366 DURÉE DÉFAILLANCE (H)

365 DURÉE INDISPO. TR. (H)

364 PUISS. INDISPONIBLE (MW)

496 MESURES PRISES

☐ 01 REMPLACEMENT TOTAL MATÉRIEL

☐ 04 MODIFICATION DU MATÉRIEL

☐ 06 CONTROLES MAT. IDENTIQUES

☐ 20 CHANGEMENT CONST. OU REF.

☐ 21 RÉPARATION PROVISOIRE

RÉDACTEUR

VÉRIFICATEUR

Figure 9 - Failure form - Reliability Data Collection System (SRDF) - French PWR units

5.2. Reliability files

Figure 9 gives an example of failure form of the Reliability Data Collection System. The SRDF follows 550 components per PWR unit (approximately 50 pumps and 250 valves). This equipment is generally connected to the safety of nuclear units. Several types of forms are produced: the descriptive card, the operation card (specifying the number of hours in service, etc.) and the failure card listing all the failure descriptors (see figure 9).

5.3. The collection problem

Data collection is a basic feature of any feedback of experience file. Is it not the reflection of the quality of data and subsequently, of the studies and analyses using these data? Collection poses the problem of training the personnel in charge as well as motivation problems.

It mainly poses the problem of a priori analysis of the failure, which often necessitates expertise, as shown in figure 10.

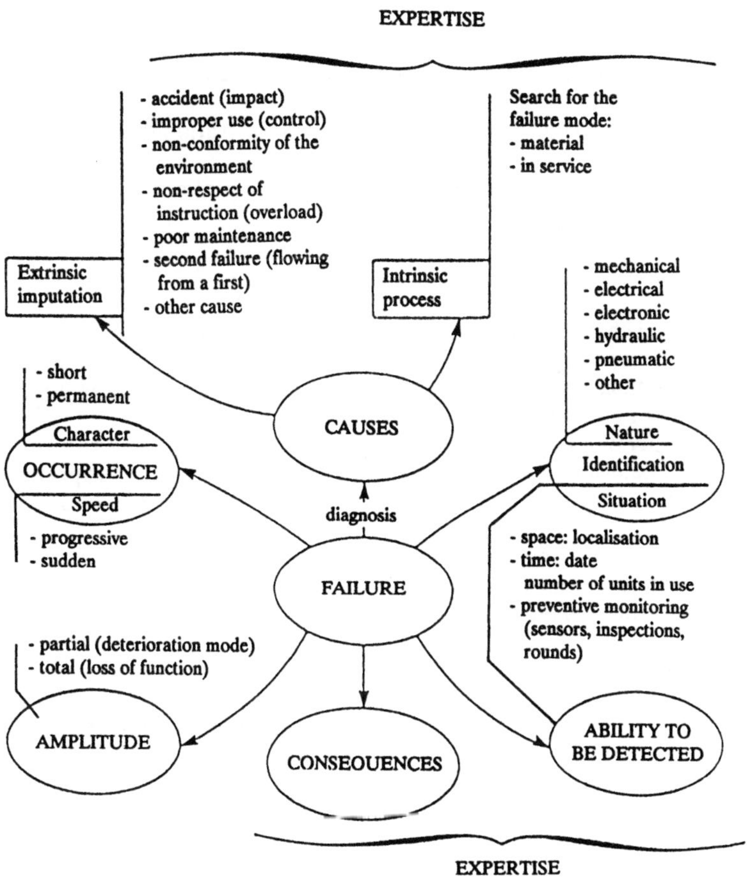

Figure 10 - List of information to be collected for drawing up a failure form

Help to the operator for collecting data is absolutely necessary. An example is given on figure 11, in the case of pumps, showing a logical failure analysis linking various descriptors of the form. Note that it is very important to define the boundaries of the component and to list all the subcomponents of this component.

Finally, it is obvious that a in situ check of collected data is necessary for warranting data quality, the in situ check being more effective than a a posteriori check performed during valuation of information just before analysis and interpretation.

6. Application - Use of feedback of experience ba

6.1. Possible processing operations

It is impossible to relate all the possible uses of feedback of experience banks in detail. This will be discussed in future papers. Notwithstanding, the main operations are:

- access to data, including the search for key-words and strings of characters,
- assistance to sorting of information,
- assessment of the quality of data,
- access to graphic programmes,
- descriptive statistical analyses,
- multivariate analyses, regression searches, survival data analysis,
- reliability computations,
- printing of compilations (handbooks of reliability data, operation data, initiating events, etc.),
- trend of characters (parameters) as a function of age or the calendar year,
- assistance to decision-making,
- comparison of performance,
- profile of operation of units,
- etc.

Some broad characteristics should nevertheless be underlined:

- the problem of fast and easy access to information, necessitating on the one hand, the use of relational database management systems and on the other, the federation of existing information on media, systems, different applications,
- the problem of quality of data: their consistency, validity, exhaustivity,
- the problem of a posteriori analysis of the failure: any feedback of experience study necessitates an often visual analysis by an expert before taking the information into account.

Finally, it is quite obvious that a in situ check is necessary for validating the quality of data, this on line control being more efficient than a a posteriori check made during valuation of information before its use.

6.2. A necessary initial processing operation: the Pareto diagrams

The present paper is limited to this processing operation which illustrates the three concepts of reliability, availability and maintainability and consequently meets the main prime objectives of data banks:

- to provide reliability parameters for safety,
- to secure knowledge of the weak points to reduce the rate of outage,
- to improve the ability for maintenance.

After having defined a functional group (system, component, etc.), the consultation of the feedback of experience bank allows computation of:

- the number of failures n (events) recorded as per functional group or their frequency of occurrence,
- average unavailability time t following these failures (or events),
- product of nt of the above two variables which corresponds to the loss of availability due to each functional group.

SCENARIO	STATE	MODES	INTERNAL COMPONENTS	CAUSES
Normal operation Test	In service Stress	External leakage of the fluid channelled	External tightness seal Internal tightness seal Coupling	Corrosion Ageing Loosening
Normal operation Test	In service Stress	Lubrication (Loss - Deterioration - Leakage)	Internal lubricating device Bearings Instrumentation	Lubrication Assembly Clogging - Obstruction
Normal operation Test	In service Stress	Coolant (Loss - Deterioration - Leakage)	Internal coolant system Internal instrumentation	Assembly Clogging - Obstruction
Normal operation Test	In service Stress	Heating	Internal tightness seal Internal coolant system Internal lubrication system Bearings Internal instrumentation	Clogging - Obstruction Lubrication Seizing Loosening
Normal operation Test	In service Stress	Characteristic (Loss - Deterioration - Cavitation)	Internal tightness seal External tightness seal Bearings Wheel - Pulley-wheel Internal instrumentation Internal drying process	Clogging - Obstruction Lubrication Seizing
Normal operation Test	In service Stress	Vibrations (and noise)	Shaft Bearings Wheel Coupling Internal balancing Device Cylinder	Regulation Assembly Balancing Wear Breakage
Normal operation Test	In service Stress	Blocking	Shaft Bearings Wheel	Breakage Seizing Clogging - Obstruction
Maintenance	Shutdown	Maintenance - Inspection		Maintenance
Maintenance	Shutdown	Modification		Modification file

Figure 11 - Logical analysis of failure - Case of pumps

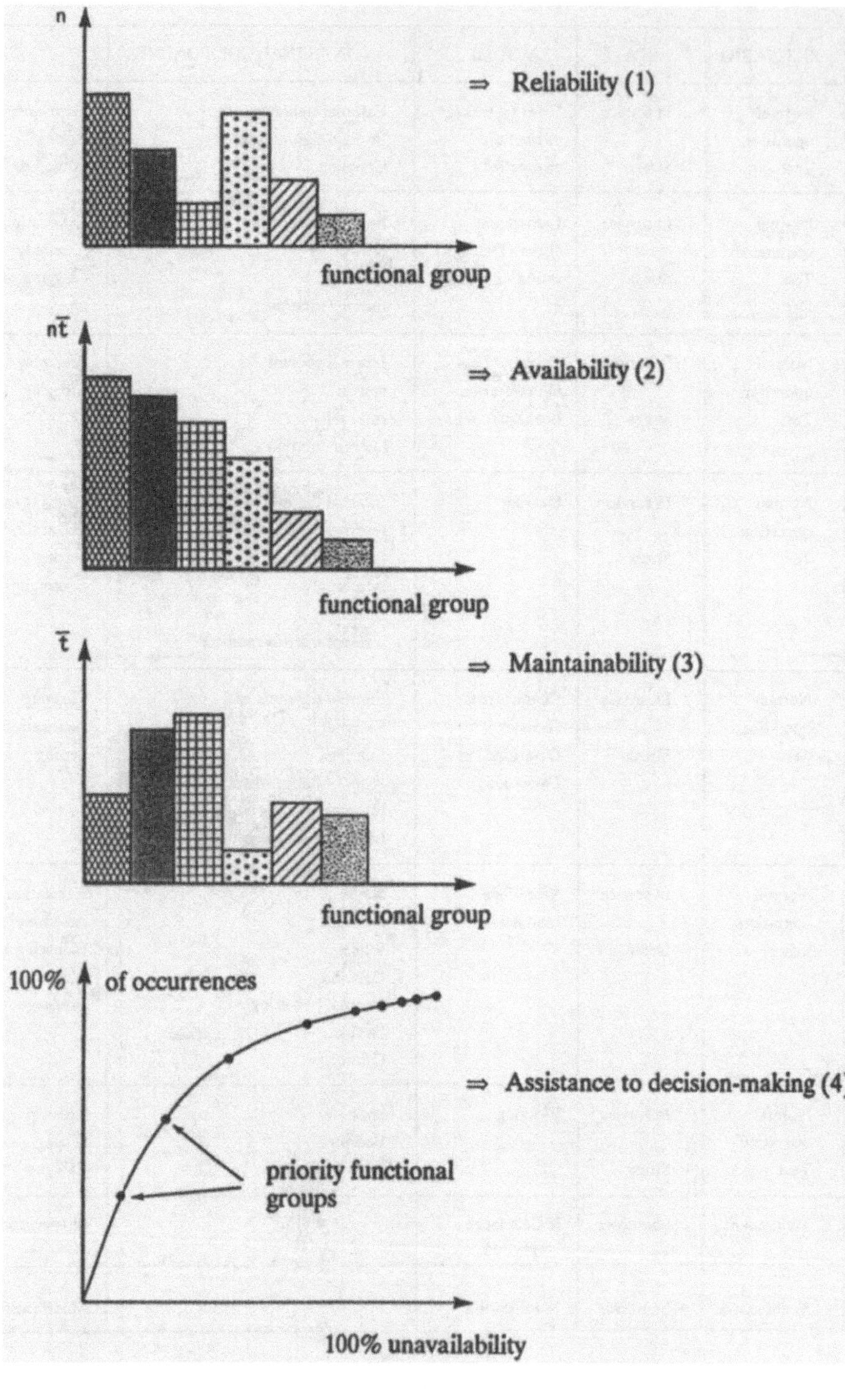

Figure 12 - A first processing operation: the Pareto diagrams

Plotting of the three diagrams allows the reliability (figure 12-1), availability (figure 12-2) and maintainability (figure 12-3) indicators to be defined.

Combined with the decision-making assistance diagram (figure 12-4), these indicators provide an aggregate analysis and are used to determine the order of priority of the actions to be conducted as they reveal the most penalising functional groups.

7. Conclusion

The purpose of this paper was:

- to specify the terminology,
- to show the interest of creating feedback of experience files and the difficulty of this organisation,
- to justify the existence of these files since their processing is a mine of teachings relating to the safety of installations and the maintenance of equipment.

The feedback of experience is one of the keys to the mastery of an installation and forestalls the risks that it may engender.

General remark: Several definitions given in this paper are excerpts from French standards. Some figures are extracted from the book "La fonction maintenance" of F. Monchy (Masson, 1987).

INVENTORY AND FAILURE DATA

T. R. MOSS
R.M. Consultants Ltd
Suite 7, Hitching Court
Abingdon Business Park
ABINGDON
Oxon, OX14 1RA

ABSTRACT. Computerised failure event data banks are employed by organisations concerned with the reliability of their plant. Inventory information on the engineering and functional features need to be stored in the bank as well as details of each failure. It is important that the information is comprehensive and coded so that the analysis of the failure data can proceed without problems. This paper discusses the basic information requirements and the procedures which need to be implemented when setting up a failure event data bank.

1. INTRODUCTION

Reliability data have many applications in safety, availability and maintenance studies. Although generic data can often be employed, at some stage there will generally be a need to collect and analyse data from specific equipment. Here the basic requirements for reliability event data collection and analysis are discussed. The examples given relate to collection and analysis of event data to provide representative parameters for RAM (Reliability, Availability, Maintainability) studies.

2. DATA COLLECTION AND PROCESSING FOR RAM STUDIES

The data requirements for RAM studies fall into two main categories:

Inventory data
Event data

Inventory Data comprises a set of information which identifies each piece of equipment by identification codes and its major design, construction, operation and process parameters.
This set of information should provide details of:

the type of equipment
where it is installed
how it was designed
how it was manufactured

J. Flamm and T. Luisi (eds.), Reliability Data Collection and Analysis, 61–71.

how it is usually maintained
the relevant process parameters

Essentially this Inventory Data set should consist of four sections:

(a) Identification Parameters
(b) Manufacturing and Design Parameters
(c) Maintenance and Test Parameters
(d) Engineering and Process Parameters

Each section may contain all, or only part of, the detailed information shown in Figure 1.

The first three sections, fields 1 to 15, constitute a set of standard information common to all the different classes of equipment installed in each facility (instruments, electrical or mechanical devices). The fourth section, fields 16 and 17, are unique for each class of equipment; they must be defined by reference to the specific design/process descriptors of each class of equipment. As an example, for the item class "Centrifugal Pumps", these specialised sections of the Inventory Data set could be:

16 Engineering Parameters
- 16.1 Body material
- 16.2 Impeller material
- 16.3 Seal type
- 16.4 Bearing type
- 16.5 Lubrication type
- 16.6 Number of stages
- 16.7 Impeller type
- 16.8 Coupling type
- 16.9 Rotating speed

17 Process Parameters
- 17.1 Flow rate
- 17.2 Suction pressure
- 17.3 Discharge pressure
- 17.4 Temperature
- 17.5 NPSH
- 17.6 Load factor
- 17.7 Media

The Inventory Data set may be stored in a special Inventory Data File either in full or in reduced form with indexes referring to other company files. For example, all or part of the design/process parameters in the specialised sections may be stored in a separate file whose address is available from the main technical data file. These separate files can be fully computerised or partially supported by hard copy of the relevant document and manufacturers drawings.

Event Data constitutes a set of information for each equipment describing the history of its operation. This history is usually composed of strings of discrete events in a time sequence, such as a "failure event string". Other typical events occuring during the life of a piece of equipment are modifications, tests, insertion into operation and shutdown from operation.

It can be seen that the operational history of an item is made of a series of event strings of the following types:

(a) Failure event string	Figure 2
(b) Changeover event string	Figure 3
(c) Replacement event string	Figure 4

In certain cases it may also include:

(d) Modification event string
(e) Test event string
(f) Insertion in operation event string
(g) Shutdown from operation event string

The Event Data Set is the structured file containing the event string descriptors for each relevant piece of equipment. This file will be fed with suitable Event Reports, that is, with forms describing each event with the relevant set of descriptors.

Thus each Event Report Form may contain all, or part of, the following set of information:

(a) Item Identification	Tag number (positional), Unique ID number (personal) and Generic code (global)
(b) Event Type	Failure, or Changeover Replacement Modification Test Insertion, or Shutdown
(c) Time Allocation	
Failure Events	Date/time of failure detection Date/time maintenance action begins Date/time maintenance action completed Date/time equipment cleared for operation Date/time equipment back in operation
Changeover Events	Date/time of changeover action
Replacement Events	Date/time of replacement
Modification Events	Date/time modification action begins Date/time modification action completed
Test Events	Date/time of test action
Insertion Events	Date/time of insertion into operation
Shutdown Events	Date/time of shutdown
(d) Event Descriptors	
Failure Events	Failure Mode Failure Cause Failure Consequences Failure Detection Mode Restoration Mode Crafts Employed
Changeover Events	Standby unit identification
Replacement Events	Replacing unit identification
Modification Events	Modification type
Insertion Events	Reason for insertion

Shutdown Events	Reason for shutdown (due to either the component or the system).

The suggested lay-out of a typical Event Report Form is shown in Figure 5.

3. SYSTEM IDENTIFICATION

Before deciding how and where to collect and store Inventory and Event Data, it is necessary to define the objective and operating philosophy of the system proposed.

For a typical system the objective could be to derive RAM parameters (failure rates, failure modes, repair rates, etc.) for selected samples of relevant components. The available sources of this information would be two major files, the Inventory Data File and the Event Data File. The link between the two files is the item identification data recorded in both files, that is, a combination of the Tag number, and the Unique Identification number, or Generic Class Code.

The selection of the relevant records from the Inventory Data File should be possible at the desired level of detail, the two extreme levels being a unique item selection (by means of Tag or Unique Identification No.) or an overall component class selection (by means of the Generic Class Code). Intermediate levels are those specifying the Generic Class Code (ie Centrifugal Pumps) plus one or more parameters of the Inventory Data Sheet (ie Manufacturer, Media, Rotational Speed etc).

The most useful tool for making such a selection will be a suitable DBMS (Data Base Management System) capable of searching the Inventory Data File by the parameters specified. Once the DBMS has identified the relevant inventory sheets, their content together with the associated event reports should then be transferred into an intermediate file for further processing. The event reports associated with the selected sample of items are identifiable via their Tag or Unique Identification number or Generic Class Code. Depending on the purpose of the analysis either all the event reports will be transferred into the intermediate file or only those having pre-defined parameters; that is, those dealing with a specified failure mode. Thus, the DBMS should be capable of searching the Event Report File at the desired level of detail for events associated with the selected inventory items. The content of the intermediate file will then be processed manually or by suitable statistical analysis programs and the relevant RAM parameters derived.
This data retrieval and processing system must be flexible, having the capability of producing either generic RAM data (ie, failure rate of centrifugal pumps) or very detailed data (ie failure rate of centrifugal pumps manufactured by (say) Worthington, on seawater service, with rotating speed up to 3,000 rpm, when the failure mode was major leakage from the seals).

The flow chart of the system proposed is shown in Figure 6. Operation of the system initially will restrict enquiries to the generic level because of limitations in the number of reports. When the contents of the Event File have expanded, more detailed enquiries become possible.

The problem is then to compare the conceptual flow diagram proposed in Figure 6, with existing or planned company organisation for operations and maintenance management.

For small event data banks it will generally be possible to restrict the amount of data collected to less than the information shown here. Nevertheless, it is important to proceed in a disciplined way so that the data generated are truly meaningful for the purpose for which they are intended.

4. PROCEDURE FOR SETTING UP AN EVENT DATA BANK

To establish a small event data bank to provide parametric data for RAM studies from the general information shown here the following steps are recommended:

1. Identify to generic classes of items on which RAM data are required
2. Define the physical boundaries of each equipment class
3. Compile lists of Tag/Unique numbers to establish populations for each generic class
4. Define minimum sample sizes and the reliability parameters to be derived
5. List the event data input to derive the required reliability parameter output
6. List the assumptions to be made in analysing the event data and the tests proposed to validate these assumptions
7. Modify the Inventory and Event Forms (Figures 1 and 5) to include only the input data required
8. Define the terms used in the Inventory and Event Forms
9. Develop Information Flow Diagrams tracing the routes for data input to reliability parameter output
10. Develop procedures for collecting and inputting Inventory and Event data into the DBMS
11. Carry out a pilot exercise to identify any problem areas
12. Modify the procedures, if necessary, and start data collection

INVENTORY DATA

IDENTIFICATION PARAMETERS

1. Tag number
2. Unique Identification number
3. Generic Class code
4. Facility/Plant identifier
5. System Identifier

MANUFACTURING AND DESIGN PARAMETERS

6. Manufacturer
7. Model
8. Date of Manufacture
9. Date of installation
10. Technical Specification Reference
11. Design code
12. Installation code

MAINTENANCE AND TEST PARAMETERS

13. Maintenance type
14. Maintenance frequency
15. Test frequency

ENGINEERING AND PROCESS PARAMETERS

16.1	Engineering parameters	
	"	
	"	eg, materials, component
	"	type, speed, etc.
16.n	"	
17.1	Process parameters	
	"	
	"	
	"	eg, pressure, flow rate,
	"	temperature, etc.
17.n		

FIG 1 INVENTORY DATA SHEET

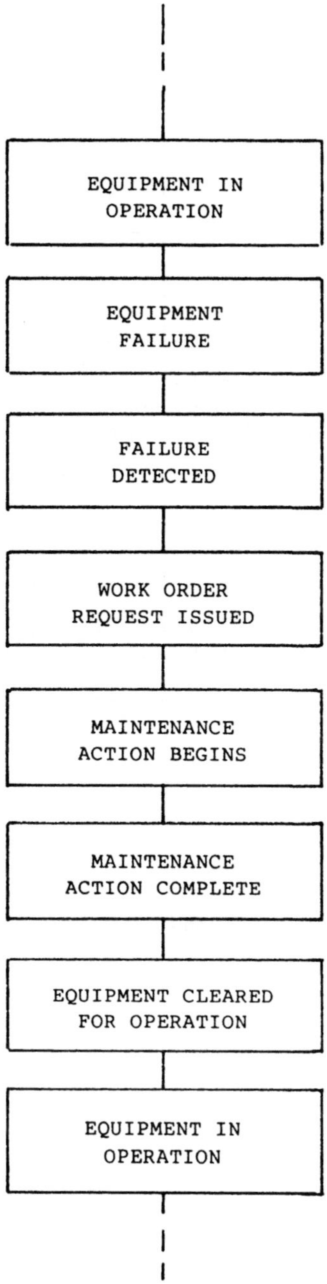

FIG 2 "FAILURE EVENT STRING"

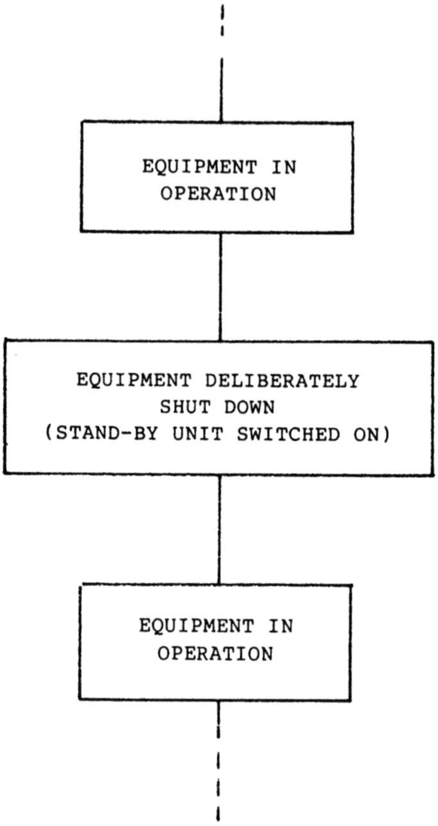

FIG.3 CHANGE-OVER STRING

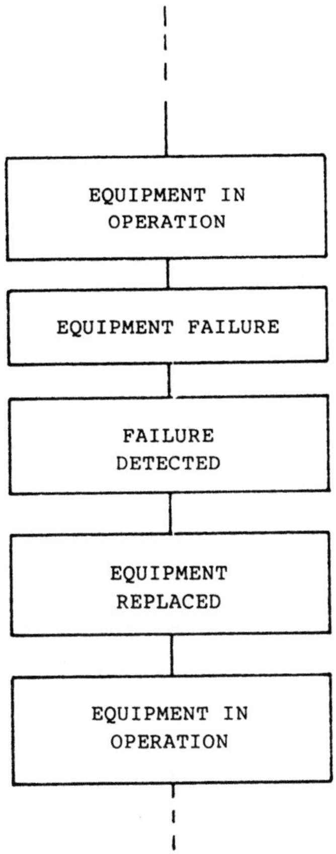

FIG.4 REPLACEMENT EVENT STRING

EVENT REPORT

ITEM IDENTIFICATION DATA:	
Tag. No. Unit ID No. Generic code	REPORT NO. COMPLETED BY: APPROVED BY: DATA:

EVENT TYPE:

TIME ALLOCATION	DATE	TIME
FAILURE DETECTION:		
START MAINT. ACTION:		
COMPLETE MAINT. ACTION:		
READY FOR OPERATION:		

FAILURE MODE

1.
2.
3.
4.
5.

EFFECT ON SYSTEM

1.
2.
3.

RESTORATION MODE	ENGINEERING CRAFT HOURS
1.	1. MECHANICAL
2.	2. ELECTRICAL
3.	3. INSTRUMENTS
4.	4. OTHERS

ENVIRONMENT	High	Normal	Low
Ambient temperature			
Humidity			
Dust			
Vibration			

EVENT DESCRIPTION: (text)

(Note other important environmental factors which could contribute to failure in Event Description.)

FIG. 5 EVENT REPORT FORM

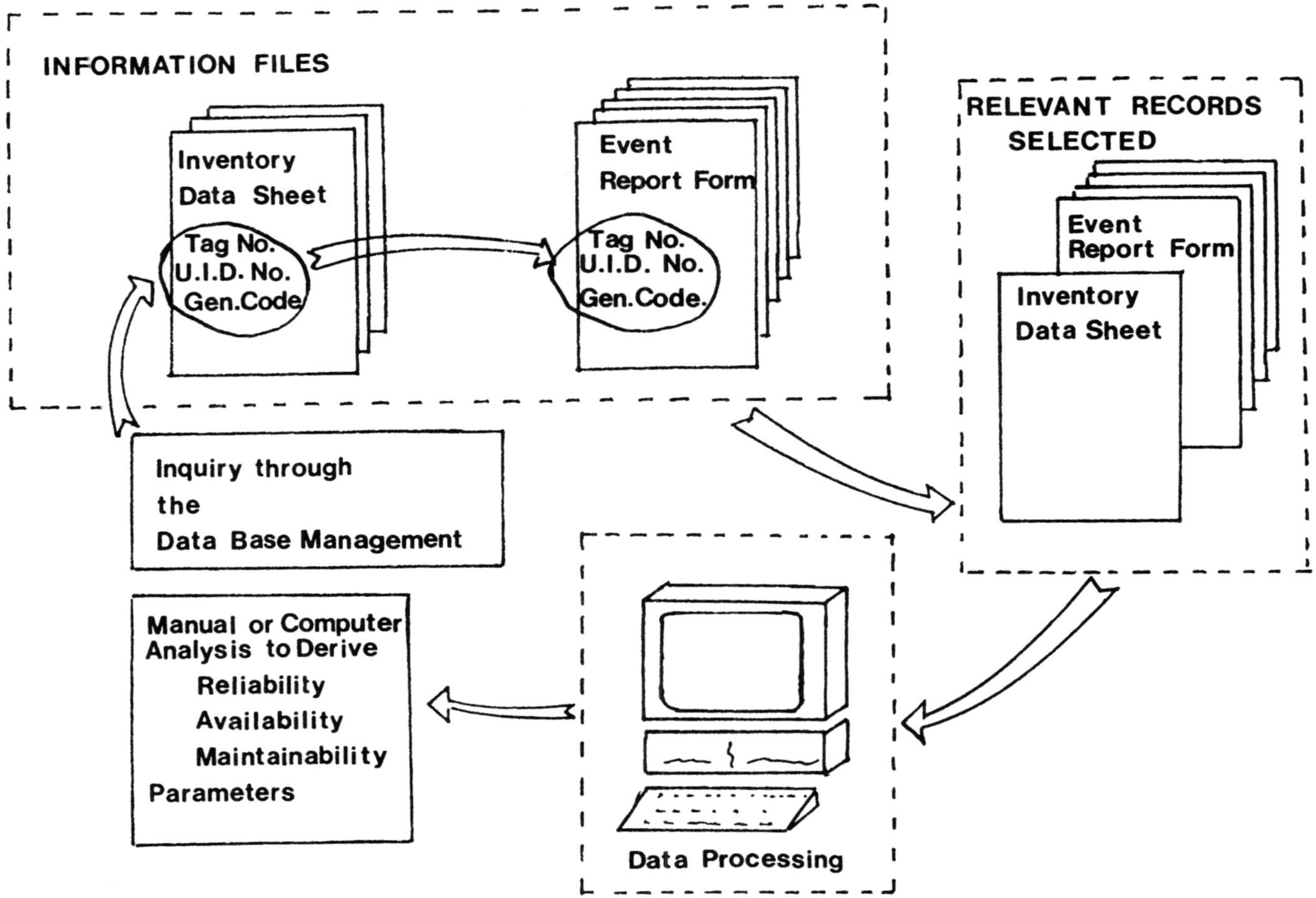

FIG.6 SYSTEM FLOWCHART

RELIABILITY DATA COLLECTION AND ITS QUALITY CONTROL

T.R. MOSS
RM Consultants Ltd, Abingdon UK

SUMMARY

Quality assurance in data collection and processing is vital if uncertanties in the derived reliability characteristics are to be minimised. This paper reviews experience in the execution of a major data collection exercise and the measures introduced to ensure that high quality reliability data output is obtained.

1. INTRODUCTION

OREDA - the Offshore Reliability Data project started in the early 1980's. Phase I involved the collection and processing of failure information from the maintenance records of a number of offshore platforms to generate global failure rates for a wide range of safety and production equipment. About 150 different types of equipment were surveyed and the results published in the OREDA Handbook.

Phase II was a more ambitious project. The objective here was to create a robust data base of inventory and failure-event information for a range of vital topside and subsea equipment. The availability of this detailed information was seen as a significant step forward from the parametric reliability data base generated in Phase I. It provides the participating companies with the facility to select populations of equipment within a well-defined envelope of manufacturer, design and functional parameters, to calculate failure rates for the different failure modes and to review individual reports of the failure/repair activities.

This paper concentrates on experience gained during data collection for Phase II. It is based mainly on experience from a typical computerised maintenance information system and highlights some

J. Flamm and T. Luisi (eds.), Reliability Data Collection and Analysis, 73–87.

general problems associated with extracting reliability data from maintenance records.

The need for comprehensive quality assurance in such a large, diverse project is also addressed. Some details of the extent of the quality control to ensure traceability of documentation and consistency in coding and data input to the data base are discussed in the paper.

2. DATA COLLECTION

The Phase II data collection exercise was controlled by a Steering Committee under the Chairmanship of Mr Hans Jorgen Grundt of Statoil. Membership of the Committee comprised representatives from the participating companies in Norway, Italy and the United Kingdon. The main contractor was Veritec - the consultancy arm of Det Norske Veritas - who were responsible for overall financial and technical control of the project. RM Consultants Ltd carried out the collection and processing of data from the UK Sector.

In total 1600 Inventory and 8300 Failure reports were generated by the data collectors. The information required for the Inventory files was divided into General Data - as shown in Fig.1 and Equipment - Specific Data. Fig.2 shows the Equipment - Specific Data required for Pumps. The universal failure data requirement is shown in Fig.3.

The approach adopted by each sub-contractor was slightly different and based on the sub-contractor experience and the flexibility of the maintenance information system (MIS). In some cases the data were partitioned into discrete sets and then transferred in blocks onto the data collection pc where it was subsequently reviewed and processed. For the majority of cases, however, the MIS was programmed to provide hardcopy output which was subsequently reviewed, the relevant data extracted on to intermediate data forms - similar to those shown in Figs.1 to 3 - and then input to the database by professional data processing staff.

This latter procedure was partly dictated by the volume of non relevant information output in addition to the required data from the MIS. The process is quite time-consuming but nevertheless has a number of advantages which may not be immediately apparent.

These include:

1. It allows experienced process engineers (who frequently lack keyboard skills) to be employed for reviewing the data.

2. Professional data processing staff are used for inputing the data ensuring low data-transfer error rates.

3. Complete manual records are available which ensure full traceability of information. These can subsequently be used for quality checking and the addition of data not available during the first pass.

3. MISSING DATA

In most cases all the required inventory data was not available from the MIS and recourse needed to be made to other information. This included microfiche records, Piping and Instrumentation Diagrams, engineering drawings and maintenance schedules. Most of the deficiencies in the failure records were associated with the lack of information in the MIS on the cause of the failure and its effects on the system. It is important to realise that repair histories are not generally designed to provide reliability data. The job cards are completed by maintenance personnel who faithfully record details of the work carried out. From this information it is necessary to deduce the cause of the failure and its likely criticality.

The situation varied from company to company and sometimes even from platform to platform but generally the data which were particularly difficult to obtain were:

Condition-monitoring information
Instrumentation details
Redundancy
Run-times
Equipment installation dates
Actual operating pressures and temperatures

It is worth noting that considerable effort was put in by the maintenance departments on each site to provide a significant amount of the missing data. Without their enthusiastic support many of the uncertainties could not have been resolved. Clearly an exercise of this magnitude cannot be successful without the full co-operation of the site management and staff.

4. QUALITY ASSURANCE

Formal quality assurance procedures were introduced by each sub-contractor at the start and actively pursued during the course of the project. Basically this involved the submission of a detailed quality plan to Main Contractor and the establishment of a Document Control Centre (DCC) into which all project documentation was stored and recorded.

During the course of the project RMC - who were responsible for over 60% of the total data collection - recorded over 400 transmittals. These ranged from general monthly financial and technical progress reports made to Main Contractor to internal transmittals to RMC staff concerning assumptions made on equipment boundaries etc. An example of a document transmittal is given in Fig.4.

The Data Collection Guidelines issued by Main Contractor required self-check and verification by the sub-contractor. This was agreed to involve sampling the various stages of data collection and recording to ensure accuracy of data transcription and interpretation. Ten percent sampling was the norm but in instances where the number of recorded failures was small 100% sampling was employed.

Samples of data recorded on the data collection forms were checked against the marked-up hardcopy output from the MIS in the initial stages. Subsequently checks were made between the OREDA program output and the data collection forms. The emphasis was on coded fields since mispellings in the free-text field were generally self-evident and left uncorrected. An example of a completed Self-Check and Verification form is shown in Fig.5.

A final quality audit of each data collection exercise was carried out by the sub-contractors QA Director and the Project Supervising Officer. One completed QA Audit Report is shows in Fig.6.

5. PROBLEM AREAS

The main problem during inventory data collection was in identifying the information required on instrumentation, maintenance and, equipment redundancy. Defining equipment boudaries was also a problem because of the differences between companies as well as differences between company-defined and OREDA boundaries. The problem can be illustrated by noting that for one company a gas-turbine driven centrifugal compressor is described by no less than 550 sub-tab numbers. From this set of numbers those items within the OREDA boundary needed to be identified so that only failures of the relevant sub-tags were recorded in the failure data base.

Problems with failure data hinged on interpreting the historical records of repair actions in terms of the failure mode and severity definitions specified in the OREDA guidelines. The equipment history listings rarely yielded any information on whether a system failure had occurred. The Condition For Work statement generally showed whether and what kind of outage was necessary for the work. However, it gave no indication of whether the work was done at a convenient outage opportunity or whether the system had to be taken out of use to deal with the problem immediately. It was shown by experience that the most efficient approach was to consider the sub-system failure first and then deduce the system effect. The procedure adopted was:

- Identify the failed sub-system, ie the sub-system which contained the failed item.
- Select the sub-system failure mode from the Euredata list.
- Consider the effect on system operation and whether system failure could result. Record the appropriate system failure mode.
- Decide and record the severity of the related system failure.

In this way the failure modes specified in the guidelines were employed throughout the data collection phase. Inevitably however differences between analysts did arise. The assumptions and interpretations made by individual analysts were thus discussed within the team and then recorded on file.

In general the problems encountered were relatively few given that reliability data were being derived from maintenance records remote from the platforms. What was certainly under-estimated was the length of time required to extract inventory information and, to a lesser degree, to interpret the historical repair data. It is clearly important that experienced process engineers are employed for extracting such data. Even then active and sympathetic participation by the local maintenance planning department is essential if complete and consistent data are to be recorded on the data base.

6. CONCLUSIONS

Extraction of high-quality reliability data from maintenance records is possible given active collaboration between the company maintenance planning department and the data collector.

With the enthusiastic and professional support afforded to the data collection teams by the companies in this exercise a comprehensive failure experience data base for the major items of equipment used offshore has been established for OREDA participants.

Even so reliability data collection and processing based on historical maintenance records has limitations. On-going data recording based on specially-designed interface programs linking the Maintenance Information System and the reliability data base have a number of advantages and in the end are likely to be significantly less expensive.

INVENTORY REPORT GENERAL INFORMATION		
*Report No.:		
*Reported by:		
*Checked by:		
*Source:		
*Installation:		
*Item name:		
*Company Tag No.:		
*Company Sub-tag Nos.:		
Unique Nos. :		
*Taxonomy Code :		
*Function:		
*Manufacturer:		
Manufacturer of Control System:		
*Model/Type:		
*Redundant Subsyst.:		
Operational Time (hours):		
*Calendar Time (hours):		
No. of Demands/starts:		
Date Installed:		
Dates of Major Replacements:		
Note: Alphanumerical fields will be free format text.		

Notes 1. Starred items indicate a manditory requirement.
2. Certain items have predefined values

FIGURE 1 - GENERAL INVENTORY DATA FORM

INVENTORY REPORT PUMP - SPECIFIC DATA		
*Type of Driver:		
*Fluid Handled:		
Fluid Characteristics:		
*Power:		
Utilization of Capacity:		
*Suction Pressure:		
*Discharge Pressure:		
Speed:		
Number of Stages:		
Body Type:		
Shaft Orientation:		
Shaft Sealing:		
Transmission type:		
Pump Coupling:		
*Environment:		
*Maintenance Program:		
*Instrumentation:		
Pump Cooling:		
Bearing:		
Bearing Support:		
Additional Info.:		
Note: Alphanumerical fields will be free format text.		

FIGURE 2 - EQUIPMENT-SPECIFIC INVENTORY DATA FORM

FAILURE EVENT REPORT (Also to be recorded on this form: all overhauls)		
*Report No.:		
*Inventory Report No.:		
*Reported by:		
*Checked by:		
Source:		
*Taxonomy Code:		
*Failure Mode, System Level:		
Subsystem Failed:		
*Failure Mode, Subsystem(s):		
Severity Class:		
Items Repaired:		
Repair Activity:		
*Failure Detected:		
Repair Time:		
Restoration Manhours		
Method of Observation:		
Additional Info.:		

Note: Alphanumerical fields will be free format text.

FIGURE 3 - FAILURE DATA FORM

RM Consultants Ltd

Suite 7 Hitching Court
Abingdon Business Park
Abingdon
Oxford OX14 1DY

DRAWING/DOCUMENT TRANSMITTAL NOTE

To T R MOSS A O F VENTON J M MORGAN A B RITCHIE J P STEAD DOCUMENT CONTROL CENTRE	Date 5/1/88 Job No. J1161 Ref. No. OR/GEN/04
Project OREDA PHASE II STUDY	

No. of copies	Reference No.	Rev.	Title
1	OR/GEN/04	1	Self check & verification guidance notes

Notes

Purpose of issue For use as guide

FROM [signature]

RM CONSULTANTS LTD ABINGDON

FIGURE 4 - SELF-CHECK AND VERIFICATION NOTES

Self check and verification guidance notes

These notes are based on a detailed examination of the Red Book and the draft Guidelines for data collection. The aim was to establish the most economical way of satisfying Veritec's requirements. A need to keep a record of assumptions is seen. It would provide a reference to assist in achieving consistency between participants and a convenient basis for self check confirmation that assumptions are relevant and consistent.

Self Check

Percentage of items prepared *
All items +

1. Task and resources - before start

   Check time and manhours allowed for each item. Is it sufficient?

2. Data check - before finalisation

Reference	*)	Check consistency of source
Consistency	*)	pattern
Interpretations	*	Interpretations must be recorded with asumptions. + Check all for consistency at recording time.
Soundness	*	Is it complete, is coverage adequate, is is sensible.

3. Assumptions

Relevance	*	Assumptions must be centrally recorded. + Check all for
Consistency	*	consistency and relevance at recording time.

4. Calculations

Arithmetic	+	When completed
	*	before finalisation.

5. Computer records and reports

   Proof reading +
   Figures +
   Graphic reliability +

Self check should be applied to any results to be delivered.
Self check at end of each part of paper phase.
Self check at end of each part of computer entry phase.

FIGURE 4a - SELF-CHECK AND VERIFICATION NOTES

Internal Verification

All deliveries except progress report, invoices, minutes of meetings.

Focus on - main conclusions
basic methods
verification of self checks.

Deliverables - Report for each equipment class.
Data bases on disk.

1. At beginning of an individuals work

Check approach

Source documents
Method of data selection
Data marking up
Data recording
Identification of missing data
Action to record, obtain missing data
Assumptions
Recording and checking record.

2. When first set of data forms are complete.

Interpretation

Check forms in association with individual to confirm.
They follow guidelines
Data sources are appropriate
Calculations, assessments and assumptions are correct.
Asumptions are recorded correctly.

3. When first platform complete
Check three sets of data forms to confirm interpretation.
Verify reports and computerised records.

FIGURE 4b - SELF-CHECK AND VERIFICATION NOTES

PLATFORM: XX SUBCONTRACTOR: RMC
OPERATOR: XYZ OFFSHORE CO. SUBJ. PROJ. MAN.: T R MOSS

	Item to be verified. GAS EXPORT COMPRESSORS Planned date. 9/2/88	Checked/Verified Date	Sign.
Self Check: - Tasks and resources	RESOURCES ARE ACCEPTABLE. IT IS NOTED THAT THE PLATFORM POSSESSES ONLY 2 COMPRESSORS AND NOT 4 AS ORIGINALLY PROPOSED.	9/2/88	J.P. Sta[illegible]
- Data	ALL FAILURE REPORTING FROM GT/COMPRESSOR UNIT RUN REPORTS. ALL INVENTORY REPORTING FROM DATA COLLECTED IN ABERDEEN.	9/2/88	J.P. Sta[illegible]
- Assumptions	ASSUMPTIONS WHERE NEEDED ARE RECORDED ON SOURCE MATERIAL AGAINST APPROPRIATE ITEM.	9/2/88	J.P. Sta[illegible]
- Calculations	CALCULATIONS - MINS/DECIMAL USING CONVERSION TABLE OF OR/GEN/28. YRS/HRS USING 1YR = 8760 HRS.	9/2/88	J.P. Sta[illegible]
- Report/Notes	REPORTING ONTO RMC/VERITEC STANDARD FORMS WITH SPECIAL ATTENTION PAID TO STARRED DATA FIELDS.	9/2/88	J.P. Sta[illegible]

Internal Verification

- Main Approach

The main approach is satisfactory.

- Reports/Notes

Data entry to the data base dis is generally satisfactory. However, the accuracy of data entry needs systematic checking.

[signature] 4/2/88

FIGURE 5 - SELF CHECK AND VERIFICATION EXAMPLE

R M CONSULTANTS

Report No.....1988.

Sheet 1 of 3

AUDIT REPORT - XX EXERCISE

Auditors	Names	Signatures	Audit Date
T R Moss A B Ritchie			23/8/88

General Comments:-

Very satisfactory audit, minor error found in data transfer from forms to database should be eliminated in final pass.

Reallocation of responsibilities transmitted verbally should be formally transmitted.

QA Director: T R Moss (for J M Morgan)

QA Manager: A B Ritchie (for J Deane)

OREDA

QA AUDIT - CHECK LIST

No.	Question	Satisfactory Yes	Satisfactory No	Remarks
	PLANNING AND CONTRACT CONTROL			
1.	Ask to see a sample of client specification (CTR's) for work undertaken. Check that the specification has been adequately reviewed to identify and provide for special or unusual requirements.	✓		
2.	Where required, has a quality plan been produced? Sample quality plan.	✓		
3.	Has Project Officer been appointed to be responsible for the work?	✓		Reallocation of Project Officer responsibilities not promulagated formally since JMM transfer to other work.

FIGURE 6 - QA AUDIT REPORT EXAMPLE

4.	Where work is shared between consultants have the interfaces been defined?	√	
	REVIEW PROJECT FILES		
5.	Is the work adequately documented, with documents correctly dated and identified in accordance with RMC P1 and P2?	√	
6.	Have the correct project control procedures been followed in accordance with RMC P3 and P4?	√	Special spreadsheet program developed and employed to meet clients requirements. Essentially same information is required in P4.
7.	Have progress meetings been held and minuted?	√	
8.	Where required, have project reviews been held?	√	
9.	Is there evidence of project planning? Sample planning charts. Are the plans kept up to date?	√	
10.	Are data collection records and reports dated and signed by the originator and where required checked and approved?	√	Some inventory reports not dated but not required on. Veritec form. Date included in database.
11.	Are Veritec Guidelines to Data Collection available? Ask to see copies. Sample self-check and verification forms.	√	
12.	Are data source references noted on Inventory and failure reports? Sample reports.	√	
13.	Are boundaries defined in accordance with guideline document? Ask to see copies of marked-up Boundary specifications.	√	
14.	Are assumptions made in applying Inventory and Failure Report definitions noted? Ask to see list of assumptions for each equipment class.	√	Original assumption noted on file. Current assumptions maintained on noticeboard.

FIGURE 6a - QA AUDIT REPORT EXAMPLE

15.	Are recommended failure mode and severity classes being strictly adhered to? Sample hard copy forms for each equipment class. Check on completeness of data recorded.	√	Some missing data - particularly trade manhours although total manhours available.
16.	Has data been transferred correctly from I and F reports to diskette? Sample diskette data and check against hardcopy.	√	Original diskette (compressors) inventory sampled. Pignone spelt wrongly. 1 error in Additional Information - further samples no errors - corrections now being done.
17.	For main contractor diskettes check that commercially sensitive data has been deleted. Sample diskette data.	x	No longer required.
	<u>REPORTS</u>		
18.	Are project reports to the agreed specification? Are reports signed by the originator and checked and approved? Sample reports.	√	Draft report to spec.

<u>FIGURE 6b - QA AUDIT REPORT EXAMPLE</u>

FACTS - A DATA BASE FOR INDUSTRIAL SAFETY

Ing. L.J.B. KOEHORST
TNO Division of Technology for Society Department of Industrial Safety

SUMMARY

Industrial accidents, special those where hazardous materials are involved, have a great impact on people and the environment. A lot of effort is spent to develop new training programs, emergency plans and risk management techniques in order to minimize the harmful effects of such accidents and to improve the industrial process safety.

In this process a lot can be learned of accidents which happened in the past. Special the experience of how to handle during an accident and the results of accident investigations can be used to improve the safety in your own situation. Important in this respect is the availability of enough and valid accident data.

For these activities our database FACTS [1] can be used. FACTS is a very large data base with worldwide information about 15,000 accidents with hazardous materials. FACTS can provide you with information about the cause, the course and the consequences of accidents. FACTS delivers computer abstracts with the most important technical details of the accidents in combination with copies of the original incident documents. In total we have over 60,000 pages of incident documents available on microfilm.

---

1) FACTS: Failure and ACcident Technical information System.

J. Flamm and T. Luisi (eds.), Reliability Data Collection and Analysis, 89–103.

FACTS has the following facilities:

- retrieval of general or very specific information about incidents; through a completely flexible search profile;
- analysis of information focussing on a variety of incident characteristics;
- identification of incident causes;
- coupling with other data bases;
- data from individual incidents can be obtained;
- copies from original incident-documents/reports can be obtained;
- abstract from undisclosed reports are available.

During the course you will get an introduction to the data base FACTS. Attention will be payed to the characteristics of the stored information, the possibilities of the retrieval programs and examples of the output. To illustrate the use of a combination of historical accident data and advanced retrieval possibilities, examples of complete analysis of historical accident data will be discussed.

Finally there will be a demonstration of FACTS. With one ore two examples the possibilities of FACTS, which have been explained, will be demonstrated.

1. Information Handling

FACTS contains information which can generally be described as: data on incidents that occured during the handling of hazardous materials. This general description implicates:

1.1 Information sources

The field in which FACTS is used is so comprehensive, that suitable information for FACTS is not limited to one single source.
The information sources used to collect incident data can be divided into five catagories.

1. Organizations with internal reporting systems. Conditions of strict anonymity apply to data derived from these sources. Police, fire brigades, labour inspectorates are the major supplying bodies.

2. Companies supply data under conditions of strict anonymity.
3. Literature and other publications annual reports, symposia etc.
4. Magazines and periodicals dealing with industrial safety, risk management, loss prevention.
5. Tidings from newspapers are used as signals for obtaining appropriate information.

It goes without saying that information from five different sources may present substantial differences. Certain sources supply more general information (newspapers), other sources such as police inspectorates give very detailed information. The time between the moment an incident occurs and publication about the incident may also vary. Publication of analysis and evaluations from major incidents sometimes takes several years. This delayed information, however, is added in FACTS to the already available information. In this way a continuous updating of information is being carried out.

1.2 Type of information

The criterious for the implementation of an event are:
- danger and/or damage to the nearby population and the environment;
- existence of acute danger of wounding and/or acute danger to property;
- accidents with hazardous materials during the following activities:
  processing,
  winning,
  transport,
  research,
  storage,
  transshipment,
  use/application,
  waste treatment
- events that can be indicated as 'near misses'.

Incidents with nuclear materials or incidents that occur during military activities will not be implemented in FACTS.

When an incident has occurred consistent with the above-mentioned incident profile, incident data are collected through the various information sources.

A discrepancy between the frequency of occuring incidents and the number of those incidents that are recorded must be accepted as a matter of fact. Incidents with minor consequences are not recorded at all.
Incidents with some consequences will be incidentally recorded. Only incidents that involve severe damage, or danger will be published, analysed and documented. The recording of the intermediate field (on a scale of seriousness) is incomplete and may depend on social relevance, industrial firms organization and other factors.

Some of these factors change from time to time.
A picture of the available information compared with what actually happens is given below.

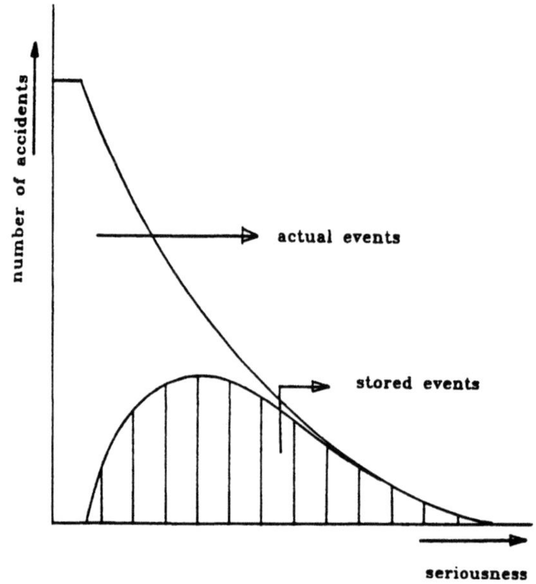

Figure 1

The available data are of a varying nature. The structure of FACTS has been chosen in such way that it must be possible to handle all these different types of information. In cases that the collected information will contain contradictions, all individual information items will be stored. No judgement is made about which is right or wrong. Interpretations based on the collected information will also not be added to this information.

In order to gain maximum profit from the collected information, high demand are being made with regard to the way information is stored.

It is important:
- to store the information in a readable way;
- to store the information thus that it becomes possible to find each piece of information;
- to have the option of adding freshly available information at any time;
- that the stored information contains one and the same data as the original information. No more, no less.

From the available information, data is used that give insight in:
- the cause of the incident;
- the course of the incident;
- the consequences to human beings, the environment and equipment.

To ensure systematic storage of information, the data are devided into several categories of keywords. With those keywords the actual coding of the incident data take place.

1.3 Model for coding incident data

A starting-point for the coding of incidents is the possibility of gaining access to the information in various ways. The original information is described and coded through the use of keywords and free text. The combination of keywords and free texts results in a summary of the original information, which is read step by step.
This model offers the possibility of coding the most variable information in an unambigious way. Each keyword can be used as a search item. Keywords may be attributes and values. The values can be considered as a subdivision of the attributes. For examples of attributes and values see figure 2, the first and the second columm. The values are hierarchically structured.

The available data are devided into a number of categories as shown in figure 2. Data referring to the course of an incident are the most important, because they indicate what actually happenend in the incident. For this purpose each incident is subdevided in a sequence of occurrences and the relevant attributes are recorded for each individual occurence. This action is the actual model for the coding of incident data. This model in question is based on a time-scale with intervals that correspond to the various occurrences that may be identified in an incident. Each occurrence often contains additional information concerning people, equipment, circumstances, technical data etc. This information is described by using the appropriate attributes with their values. If the correct value is not available or not precise enough, free text may also be used.

The actual recording of each accident is carried out in a number of lines (figure 2). The number of lines is related to the amount of available data. The first column indicates type (= attribute) of information. The second column contains the values that give more detailed information about the part the attribute is referring to. The number of attributes and values is large, about 1500, but limited. In order to allow for greater specificity, the use of free text is also permitted. This constitutes the third column, which also contains the numerical data.

2. Cause Classification

Accidents are often analysed for the purpose of determining their cause in order to try to avoid a repetition of the accident.
Several coding systems for tracing the cause of an accident have been developed. The method used in FACTS will now explained.

As has already been mentioned, in FACTS the course of an accident, is translated into a chain of occurrences. This particular chain of occurences makes it possible to describe the different actions that took place during the course of the incident. An action can be considered to constitute a single isolated part of an incident. In this way the dynamic information is described. Examples of actions which may be classified during an incident are:
natural occurrence - earth quake
incorrect operation - overspeed
abnormal condition - overheat

FACTS Database for Industrial Safety Accident abstract	Acc.#: 305

Identification:

	FILM#	REPORT	1980 001 1051 TO 1052
	PROGR	LBB	1979 0906
	SOURCE		LIT
	SDESCR		AMMONIA PLANT SAFETY VOL. 12
	ADRES	F	LIEVIN (PAS DE CALAIS)
	ADATE	1968	1968 0821
	ACTIV	TRANSSHIPMENT	FROM ROAD TO STORE
	LOCTN	FACTORY-YARD	-
	DTYPE	CHEM-INDUSTRY	-
	DCHEM	AMMONIA (PROD.)	-
	ENCIR	TEMPERATURE	22.00/C

Cause:

	CAUSE	TECHNICAL-FAILURE	STRESS CORROSION

Accident description:

1.	OCCUR	FATIQUE	BY NOT SUPPORTED OVERHANG
2.	OCCUR	CORROSION	STRESS
3.	OCCUR	CRACK	CIRCUMFERENCE
	EQINV	WELD	-
4.	OCCUR	BURST/RUPTURE	DURING UNLOADING
	EQINV	TANK	-
	EQAPPL		CONSTRUCTED FOR PROPANE AND
	EQINV	TANKVEHICLE	IN 1967 TESTED FOR AMMONIA
	QCONT		38.00/M3
	SPILL	POLLUTION	1.900E+04/KG
	CHEM	UN-1005	AMMONIA
	STATE	LIQ-GAS-PRESS	-
	TEXT		LOADING PERMIT 0.53/KG/LTR
	EQMADE		T1 STEEL
	EQDATE		1964
5.	OCCUR	RELEASE	-
	PRESS		1206.58/KPA
6.	OCCUR	VAPORIZE	-
	EQINV	VAPOUR-CLOUD	MUSHROOM SHAPED
7.	OCCUR	BLOW-AWAY	-
	EQINV	FRAGMENT	BOTH PARTS OF THE TANK
	EQINV	TANK	COLLIDED WITH HEAVY OBSTACLE
8.	OCCUR	EVACUATION	-
	HMINV	CITIZEN	IN NEIGHBOURING STREETS
	FATALS		6, 5 WORKERS AND THE DRIVER
	INJURS		15
	WNDNG	TOXIC-INHALATION	BURNS OF RESPIRATORY ORGANS
	EQACT		T1 STEEL PROHIBITED FOR STORAGE
	EQACT		AND TRANSPORT OF AMMONIA

Summary:

	SCENE		RUPTURE TANK OF TANKVEHICLE

Figure 2, example of an accident abstract.

electrical failure - power breakdown
mechanical failure - break/burst
etc.

Many causes may be identified during incident analysis. Or more precisely, every single action in an incident has at least one cause. Each action, together with a number of factors causes the next action. These factors are the values used in FACTS.

The following flowchart indicates the cause of an incident indicated.

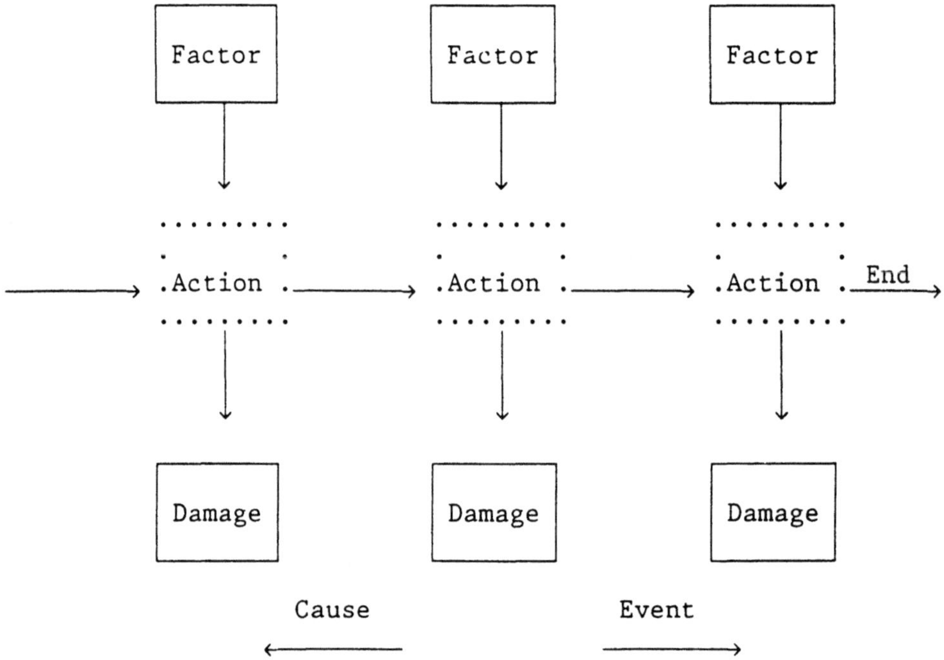

Each next action in an incident takes place as the result of the previous action. This point of view makes it possible to define one single cause of an incident without referring to certain aspects or elements in, the incident.

Definition: The cause of an incident is that cause which is related to a certain characteristic of an incident.

In accordance with the above a 'near miss' is an incident during which this particular characteristic did not occur, but reasonably have been expected to do so.

3. Database Structure

The database FACTS uses a HP-3000 system. FACTS was designed with the data base management packet IMAGE. Starting-point is a 'network' structure. This structure is build up from one detail data set and five master sets. A master contains information concerning one specific aspect (for example accident number or value numbers). The detail data set indicates the relationship between the variety of information from the several masters.

Flow scheme of the structure of FACTS.

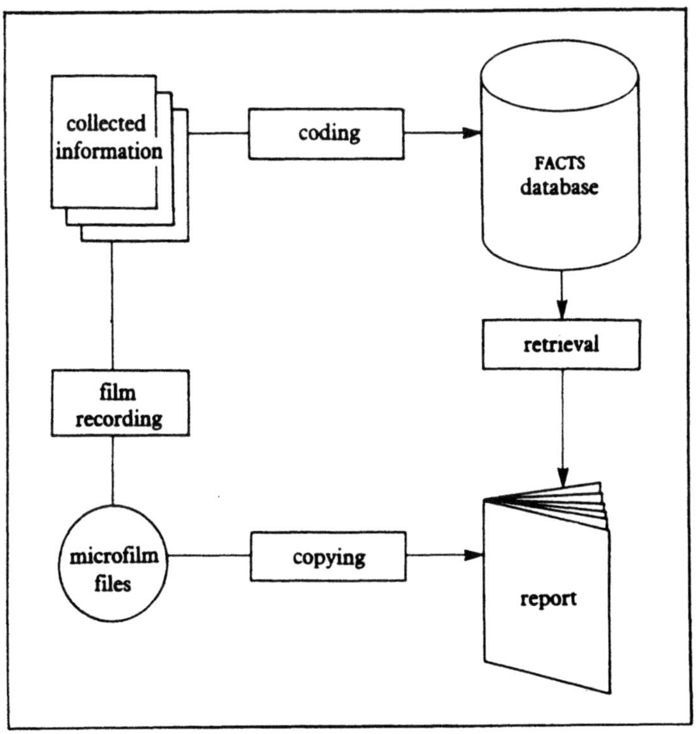

FACTS updating and service

3.1 Retrieval program FAST

To search and to prepare information from FACTS a special retrieval program (FAST) with different functions has been developed. The structure of this program is comparable with the HP-program QUERY.

The results of retrieval can be stored in a select file. Ten select files are available. Several manipulations can be made executed with these files.
When FAST is used to carry out a retrieval, the program reckons with the confidentiality of the used information.
FAST contains the following functions:

Search functions

These functions search for certain attributes, values or dimension numbers. Dimension numbers are for example UN-numbers.
The following commands are possible:
GET : search the entire data base for incidents with a certain key number (attribute, value or dimension). Controle and corrections are executed in the event that a key number appears more than once in one incident.
This avoids double-counting.

SHOW: counts how many times a key number is used in the data base.

READ: reads series numbers stored in a file. The meaning of these numbers depends on the accompanying key number, which must be defined. READ makes it possible to connect other data bases to FACTS.

Compatible functions

These functions search in FACTS. Simultaneously, several manipulations can be executed with the obtained data. Four different manupulations are possible.

1. SEARCH ADD fire tank
   Search those incidents containing fire
   and those which containing tank, and
   combine the two series.

2. SEARCH COMPARE fire tank
   Search only those incidents containing both fire and tank.

3. SEARCH REST fire tank
   Search those incidents containing fire. Those incidents that also contain tank are delected.

4. SEARCH COUNT SEL 1 fire
   Counts in how many incidents, stored in select file 1 fire are used.

4. Storage of Original Documents

As has already been mentioned, there are five information sources for FACTS. After the information from these sources has been introduced into FACTS, the original information is stored, and remains available. It may be very usefull to have the disposal of this information in combination with the coded information. A prerequisite is that the accessibility of this information, in combination with the coded information, will be very good.
It is not a very attractive option to store the original information in the shape of documents. All the original documents are recorded on microfilm. During recording each microfilm copie is given an unique filmnumber. When the information is being coded this filmnumber is added to the incident information, which will be stored in FACTS. In this way there is a direct link between the microfilm archives and FACTS. The microfilm device consist of a reader unit and a printer unit, which produces hard copies of the microfilm pictures.
The microfilm reader-printer has an interface. With this interface a link is made between FACTS and the microfilm device. The retrieval program FAST makes it possible to edit a file which contains the required filmnumbers. A specially developed control program reads the filmnumbers and activates and controls operations of the microfilm reader-printer.

4.1 Abstracts of confidential information

The confidential incident documentation is stored on microfilm, but this information is not available for external use. In order to obtain the benefits of this information, abstracts have been made.

The relevant information is written down in such a way that anonymity is guaranteed and a maximum of relevant data is represented. In this way it is possible to augment the open information with valuable information.

5. Latest Developments

After 10 years of experience with FACTS, in its present state, it was necesarry to make use of new techniques and develop new possibilities which meets the expectations of our clients. The following developments are, or will be realized in the nearest future.

- Upgrade version of FACTS on a micro VAX based on the relational DBMS ORACLE.

- Special version of FACTS on a PC, (see 5.1).

- A two days workshop, using FACTS to investigate and analyse incident information.

- A user friendly interface FIFE to support external users.

5.1 PC-FACTS

The TNO department of Industrial Safety has developed a PC version of FACTS called PC-FACTS for use as an in-house system. Besides the other services of FACTS focused on information supplying for incidental cases, PC-FACTS is specially developed for those who make regular and advanced use of historical accident data. PC-FACTS is an inhouse system and runs on the PC of our clients, offering a flexible and advanced tools at their possession to handle and analyze data of industrial accidents with hazardous materials.

<u>Available accident information</u>

PC-FACTS contains accident abstracts (see figure 2) which

gives the most import details of the accidents, described in a chronological way. More detailed information can be obtained on request from our microfilm archive.

Structure of PC-FACTS

PC-FACTS is a menu driven database. The special developed software allows the user to handle the stored accident information in a flexible way. PC-FACTS is developed in a modular way and new functions can easily added to the system. The following standard functions are available:

- Complex search facilities. With the help of an internal library of all the available keywords, search profiles can be made using .OR., .AND., or .EXCLUDE. combinations. The results of a search is stored in a selection file. Further detailed searches using these selection files are possible.
- Free text search. Besides the key words the accident abstract contain free text. String search is possible for this free text.
- Selection administration to verify existing selction files and used search profiles.
- Sort routines to sort accident records in a specific sequence.
- Edit and view functions. Accident stored in selection files can be viewed and edited.
- Print facilities to generate lists of the selected accidents according to your own specifications or to print accident abstracts.
- On-line help instructions, each menu and sub menu contains a detailed help instructions.

Besides these basic functions the following functions are optional:

- Adding new records, this can be annual updated records which will be supplied by TNO or your own company records of incidents.
- Accident analysis, the available accident abstracts can be used to investigate the chain of occurrences which occurred during an accident.
- Graphical presentation to generate pie charts and bar graphics, (in development).

PC-FACTS is available as a shell with the standard options. The shell can be used by a company to registrate and to handle their internal accident information. A detailed manual will help the users to gain the maximum results. Depending on the criteria of the users, datasets of accidents abstracts will be selected from FACTS and installed in PC-FACTS. An annual up-date, based on the same criteria is possible to keep PC-FACTS up to date.

Datasets

- accidents which occurred during one of the following main industrial activities:

		Number of accidents
- storage		1158
- transsshipment		1010
- processing		1855
- transport	* road	1090
	* rail	593
	* pipe	965
	* inland waterways	338
	* sea	418
- handling and use		1746

- accidents where one of the following chemicals were involved:

- chlorine	357
- ammonia	209
- natural gas	821
- oil, several types	> 2000
- propane	558
- LPG	735
- hydro chloric acid	160

- accidents where a specific piece of equipment or unit is involved:

- pipeline, lines	1580
- tank (storage, transport, etc.)	2100
- refinery	509

The above mentioned examples illustrates what types of data-sets are possible. In addition to these examples it is possible to create other datasets according to the criteria of our clients.

Hardware requirements

IBM-compatible PC with a hard disc of at least 20 Mbyte. The PC should be equipped with an EGA or a VGA screen.

Software requirements

PC-FACTS is developed with Dbase 3+ and compiled with Clipper. This means that except a MS-DOS version of 3.3. or higher, no additional software is necessary.

Availability

PC-FACTS is available to all type of clients or organizations. The use of the database and the installed data is strictly limited to the owner of the system. Data or software may not be selled through to third parties. The prices of PC-FACTS, and other services of FACTS are mentioned in the FACTS price information bulletin.

6. References

[1] Wingender, H.J.
Reliability Data Collection and Use in Risk and Availability Assessment.
Springer-Verlag, 1986.

[2] Bockholts, P.
A databank for industrial safety.
Seminar Industrial Safety Euredata-TNO, 1981.

[3] Bockholts, P.; Heidebrink, I.; Moss, T.R.; Butler, J.A; Fiorentini, C.; Bello, G.C.

[4] Koehorst, L.J.B.; Bockholts, P.
FACTS, Most comprehensive information system for industrial safety.
TNO, March 1987.

[5] Professional Accident Investigations, Methods and Techniques.
Institute Press, 1977.

[6] Accident Prevention Manual for Industrial Operations.
National Safety Council, 1974 (7th edition).

RELIABILITY DATA COLLECTION SYSTEM IN THE TELECOMMUNICATION FIELD

N. GARNIER
Centre National d'Etudes des Télécommunications
B.P. 40
22301 LANNION
France

1. INTRODUCTION

The Reliability Data Banks (RDBs) are organizations which collect experimental results on the time-dependent behaviour of relevant components (or considered as such). Their task is a vital one. As a matter of fact, the value of the estimation made by the "customers" depends on the quality of the numerical data supplied by the bank. The task is very complex and difficult, going well beyond a simple data collection problem. Because of the statistical nature of reliability data, the RDBs have to collect a great amount of information but they must also make sure that this information is correct. It is up to the banks to determine how to obtain data on components under real working conditions. They also have to investigate convenient tests for laboratory studies. Moreover they must interpret the primary information to process the final data sheets in tabular or graphical form in a Reliability Data Handbook.

The reliability Data Banks are generally specialized in different fields (electronics, electricity, mechanisms).

This paper presents CNET reliability data bank

2. CNET BANK

The CNET bank specializes in the field of electronic components.

For such components, both the "burn-in" and the "wear out" periods can be considered as negligible. On the other hand, assuming that both classes of failures occur at random intervals and that the number of failures is the same for equal operating times, the reliability of an electronic

J. Flamm and T. Luisi (eds.), Reliability Data Collection and Analysis, 105–124.

component is mathematically defined by the well-known exponential formula :

$$R(t) = \exp(-\lambda t)$$

where λ is a constant, the so-called failure rate. The reliability of electronic components must be defined in terms of failure rate, for internal and external conditions.

To find out the probability of an event to occur, statistically significant data must be compiled. Two methods can be used which both give good results. The first one involves laboratory reliability measurements, with as accurate a simulation as possible of the overall stress spectrum for environmental and also internal stresses. This method is costly, time-consuming and requires using a large sample. But the operating time can sometimes be reduced by a factor of up to 100, by submitting the components to a higher stress level. The second method depends on the obtainment of suitable information by observing components in actual use (in the real equipment). When the operating time has been exactly measured over extended periods of observation and all failures have been carefully noted,satisfactory reliability data can be obtained. Telecommunications plants are well adapted to the second method. Many pieces of equipment are fitted with numerous identical components operating 24 hours a day, hence a very significant product n x t can be obtained without any assumptions (n = number of failures ; t = operating time) in a relatively short time. The reliability of conventional components such as resistors, capacitors and transistors is now known with accuracy. The reliability characteristics of new integrated circuits using recent technologies of ever growing complexity or optoelectronics components are not so well-known, so the CNET has devoted all its efforts to this field. The CNET is also collecting a large amount of information from French Administrations, public services and industries to update the CNET's Reliability Data Handbook.

3. RELIABILITY DATA COLLECTION SYSTEM

3.1. HISTORY

In 1956, the French PTT Administration began to systematically control the quality of the equipment constituting its Telecommunications network.

Since that time, the procedure used to collect information on failures and also the system used to process this data have undergone successive evolutions. In view of the large number and wide variety of different equipment to control and the resulting volume of data to be processed, computerized methods have been implemented since 1965. The computerized data base system SADE (Field failure analysis system) was developed in 1976 and has been operational since 1978. The first-generation SADE system will be replaced by the second one currently under study, at the end of 1990.

3.2. POSITION OF THE PTT ADMINISTRATION

The Administration is particularly well placed to control the reliability of its equipment ; it is responsible for both operating and maintaining the network.

All malfunctions are recorded at all stages in the equipment's lifetime :

- during factory quality testing by the "Telecommunications Technical Verification Department"
- when the equipment is set up by the PTT or a technician from a private company
- throughout the equipment's operating lifetime.

3.3. SCOPE OF THE DATA BASE

The data base provides the means for analysing failures affecting a wide variety of equipment, including :

- analog and digital transmission equipment for cable and radio links
- power equipment
- electronic switching systems such as E10, AXE, MT25...

By compiling fault data (collected using the "REPAR 2000" repair form and the "REBUT 2000" for rejected active components) and data describing the equipment (mainly their population and nomenclature), it is possible (see figure 1)

- to evaluate equipment
- to output equipment maintainability statistics
- to evaluate the reliability of components used in the various systems and update the CNET's RELIABILITY DATA HANDBOOK (a new version will soon be published).

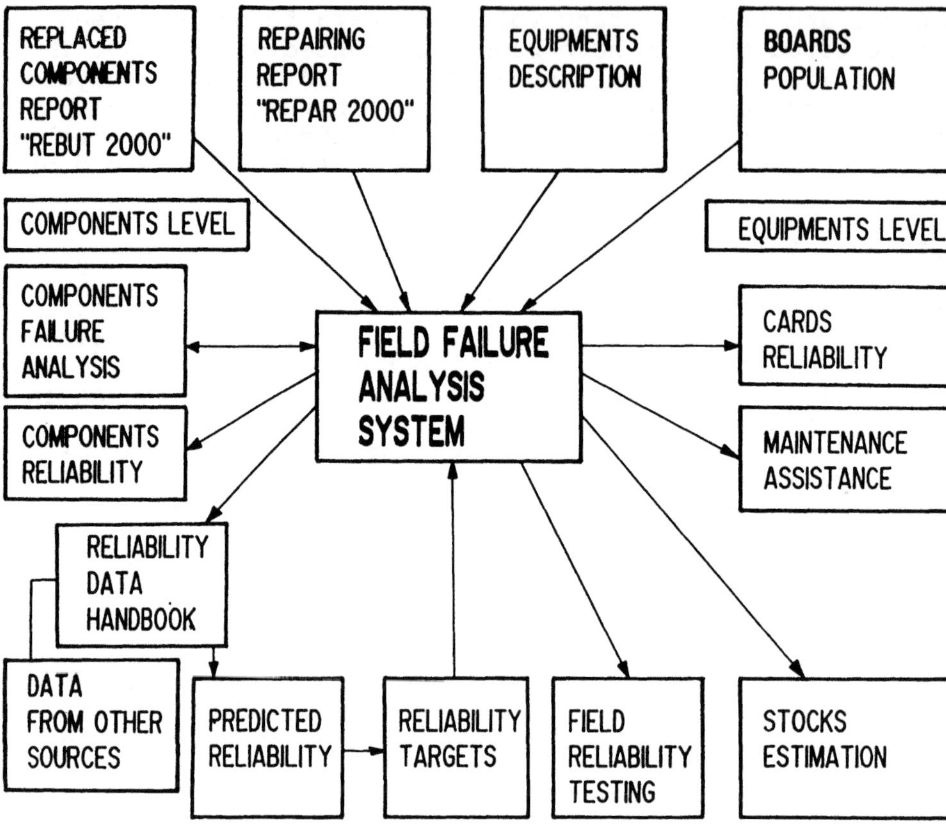

Fig. 1 - Field failure analysis system

3.4. EQUIPMENT CHECKING FORMS

Equipment is currently checked by a unique procedure based on the use of two forms : the repair form "REPAR 2000" and the rejection form "REBUT 2000".

However the requested information and the destinations of these forms vary, depending on operational constraints affecting the various equipment groups.

3.4.1. *Repair form "REPAR 2000".* This form (figure 2) is drawn up in four copies. It is divided into three parts as follows :

- the upper part is filled in when the failure occurs. The requested information mainly concerns the identity of the department reporting the failure, the various references of the failed equipment and details on the nature of the fault.
- the middle part is used for administrative details concerning the repair.
- the lower part is filled in by the repairer once the equipment has been repaired ; the elements found to be fautly are mentioned and the components which caused the fault are identified.

The copies of the form each have a particular utilization (figure 3)

- the pink copy forms the counterfoil,
- the white copy is attached to the equipment during repair, wherever this may be carried out. When it is completely filled, it is sent to the responsibles for the data base system (D.T.R.N. Malakoff for transmission equipment and M.A.T. L'Isle d'Abeau for electronic switching systems),
- the blue copy is attached to the equipment during repair and is kept by the repairer,
- the yellow copy is attached to the equipment during repair and is returned with the repaired equipment.

The repair form is drawn up in case of :

- complete failure
- intermittent failure
- drop in performance

either when the equipment is set up or during its working lifetime, whatever the type of failure may be.

3.4.2. *Rejection form "REBUT 2000".* This form (figure 4) used for systematic collection of replaced active components.

The physical and electrical analysis of the components allows identification of failure causes.

This form is filled in by the repairer.

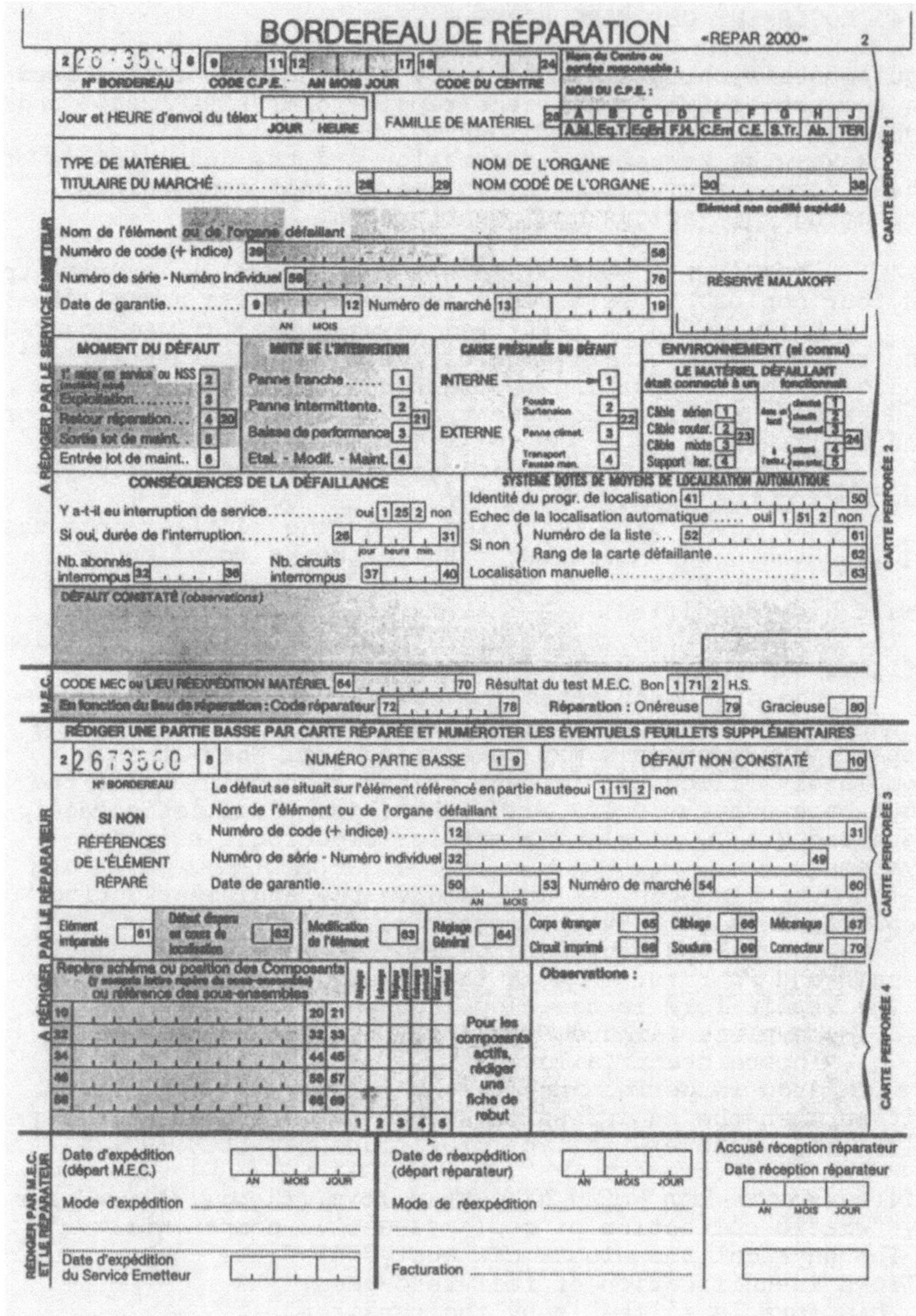

Fig. 2 - Repair form "REPAR 2000"

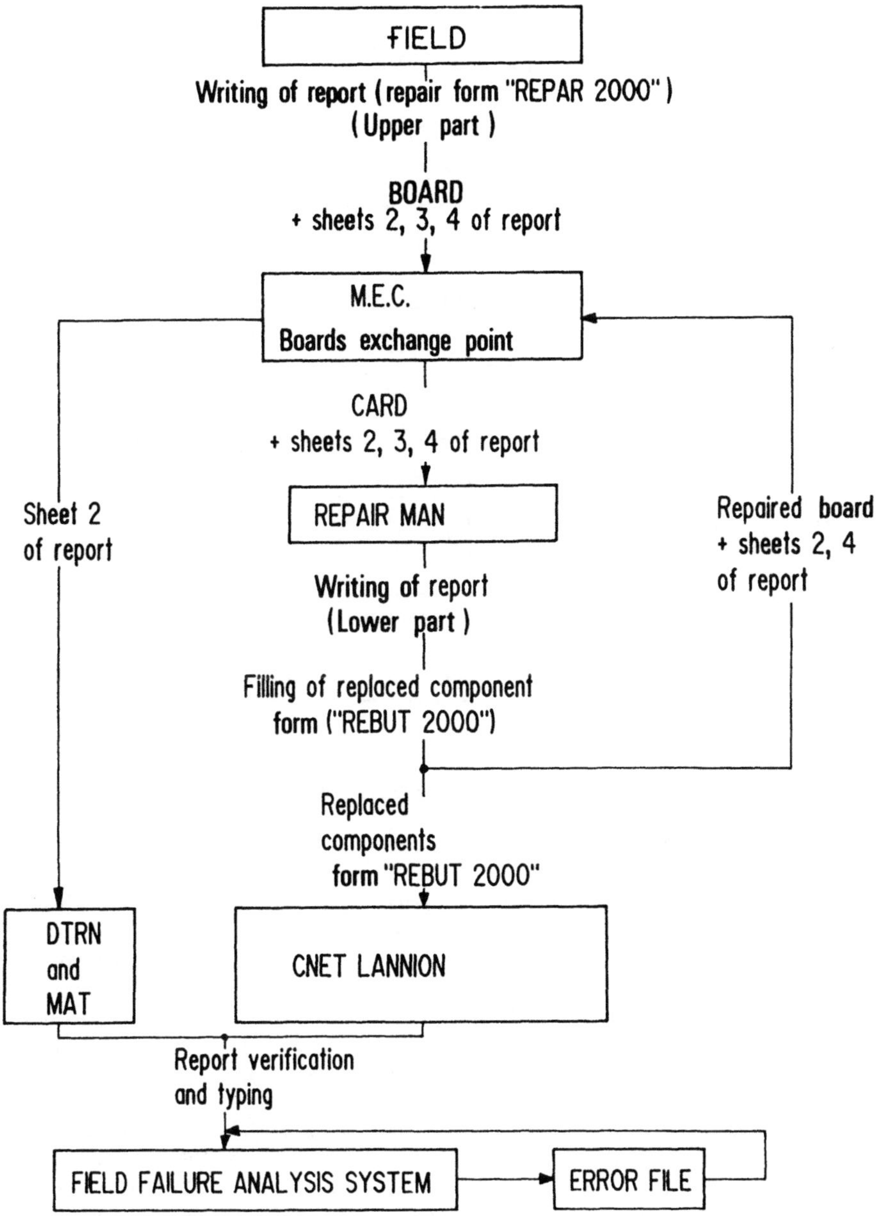

Fig. 3 : Utilization of repair and rejection forms

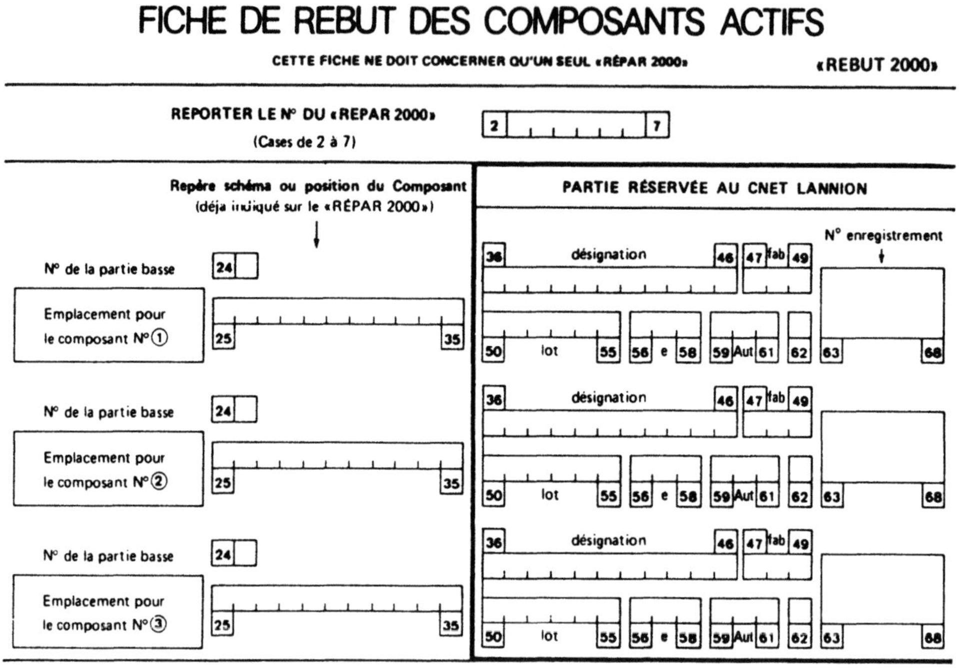

FICHE DE REBUT DES COMPOSANTS ACTIFS

CETTE FICHE NE DOIT CONCERNER QU'UN SEUL «RÉPAR 2000» «REBUT 2000»

REPORTER LE N° DU «REPAR 2000» (Cases de 2 à 7) 2 7

Repère schéma ou position du Composant (déjà indiqué sur le «RÉPAR 2000»)

PARTIE RÉSERVÉE AU CNET LANNION

N° de la partie basse 24

Emplacement pour le composant N°① 25 35

36 désignation 46 47 fab 49 N° enregistrement

50 lot 55 56 e 58 59 Aut 61 62 63 68

N° de la partie basse 24

Emplacement pour le composant N°② 25 35

36 désignation 46 47 fab 49

50 lot 55 56 e 58 59 Aut 61 62 63 68

N° de la partie basse 24

Emplacement pour le composant N°③ 25 35

36 désignation 46 47 fab 49

50 lot 55 56 e 58 59 Aut 61 62 63 68

Fig. 4 - Rejection form "REBUT 2000"

3.5. AMOUNT OF EQUIPMENT

The confidence which may be placed in the data provided by the follow-up of equipment in the field depends on the quality of the data collection process and also on the amount of equipment installed in the network.

For example, the population of electronic boards observed during one year (1/7/87-30/6/88) is around 9 290 000 :

- electronic switching systems (83 %)
- transmission equipment (17 %).

During the same period, some 180 000 failure reports were processed by the SADE system (89 % and 11 % respectively).

The examples give an idea of the quantity of data available. They can be used in laboratory tests on components, which enables the failure rates of most of them to be verified.

This accumulation of data provides information which is fundamental for relaibility studies on functional units, boards or components.

4. THE COMPUTER-CONTROLLED DATA BASE SYSTEM

4.1. SADE-1G

The data base software known as CLIO is loaded on two minicomputers (SOLAR 16/75) : one is devoted to transmission equipment, the other one to electronic switching systems (data on power equipment is available on both minicomputers).

Without going into all the details, CLIO allows the use of a network structure for the description of data : it incorporates its own data handling language (update, interrogation, printout). When CLIO is memory-resident, several users can work simultaneously.

Data, data structure and programs are stored on discs. Users simply call the required program and supply the necessary parameters.

4.1.1. *Data.* The data managed by the data base system mainly consist of the "REPAR 2000" repair forms.

Data may be entered either centrally by people responsible for the system or remotely via a computer terminal on the site when a fault has been detected.

Other data sources enable the updating of :

- the equipment list
- the equipment nomenclature description
- the populations of the various equipment groups
- the list of PTT centers.

4.1.2. *Data organization.* The adopted data organization is the result of a long concertation with system users. This organization is fundamental for the performance of data inquiry and updating.

In theory, the data structure should be independent of the programs and allow any type of application. In practice, we must find a compromise between :

- a high performance level due to a large number of rapid access points whilst maintaining data centralization.
- disc space requirements, reducing data redundancy.

Furthermore, the computerized data base system must fulfil one function which is not required of all data bases : it must be able to output summary tables which are synthetic views of the data from various angles. The outputs correspond to the principal operational programs used in the system.

4.2. SADE-2G

The second generation SADE project has been divided into two sub-systems (figure 5) :

- the "data management" sub-system, which was carried out on the central site in Toulouse, for all the equipment (transmission and switching), includes the repair, data acquisition and restoration functions.
- the "data-analysis" sub-system, which was carried out on micro-computers and includes control functions concerning equipment reliability, repair quality, stock levels, component reliability and obsolete components.

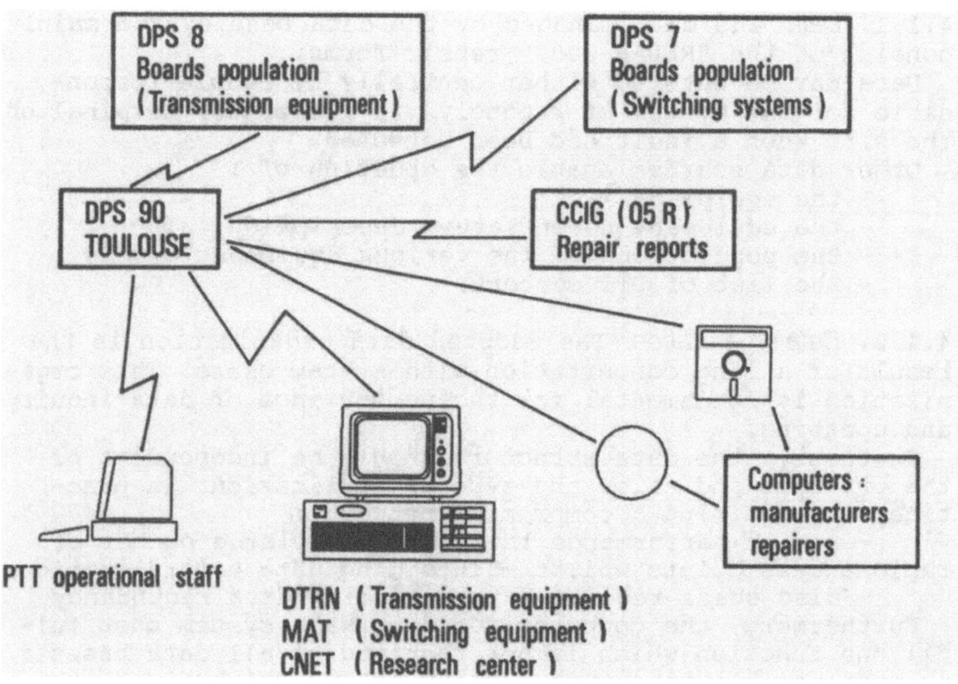

Fig. 5 - Schematic diagram

The computing equipment needed on the main site are the DPS90 of the D.T.R.N. in Toulouse with standard BULL software and four remote workstations fitted with 386 micro computers using the MULTIDIM data analysis package and the usual work processing, graphic and spread-sheet packages.

4.3. CONTROL METHOD

Equipment can be controlled in one or two different ways :

4.3.1. *Overall control of a type of equipment.* This is the most common procedure. It consists in recording all the faults affecting a given type of equipment.

The method is particularly well adapted to control equipment of which several thousand units are installed in the network. These units may have different technical features and be supplied by different manufacturers.

The method provides a means for observing the overall quality of the equipment independently of the manufacturer. Transmission and switching equipment is controlled in this way.

4.3.2. *Control by sampling.* This procedure is applied to equipment, of which several hundred of thousand of units from the same manufacture are installed in the network.

The method consists in recording only faults occuring in a clearly defined sample which is representative of the entire population ; overall equipment quality is estimated by extrapolating the results obtained.

However, the definition of the sample to be observed and the way in which the information concerning this sample is collected, must be precisely stated in instructions given to operational departments.

5. APPLICATIONS OF THE DATA BASE SYSTEM

A large number of investigations may be carried out using the data collected by the computer-controlled data base system.

5.1. RELIABILITY

The data relating to faults occuring in operating equipment can be used to determine failure rates in equipment and in components. Information contained in failure reports and that concerning analysed components in rejection forms is used to quantify the following parameters :

- inherent component reliability
- component implementation
- non-respect of standard working practice
- external constraints (lightning, power surge...)
- mishandling.

Following the evolution of the number of replacements per month per 1000 subscribers for each kind of exchange over several years can be an example of equipment reliability assessment (figure 6).

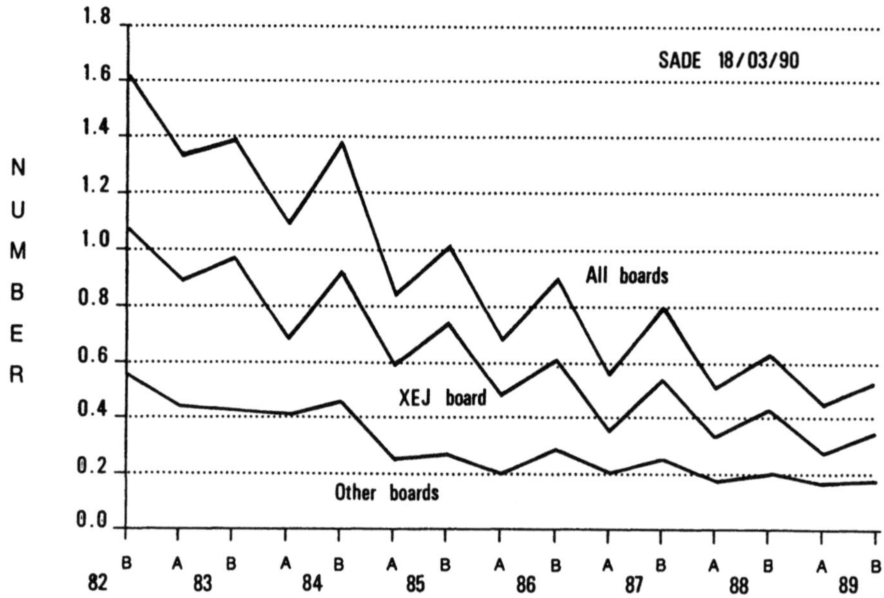

Fig. 6 - Evolution of telephone exchange reliability since 1982 (Number of replacements per month per 1000 subscribers)

5.2. MAINTAINABILITY

The repair information entered on the repair forms can be used to output a "failure tree" for each type of equipment.

5.3. ENVIRONMENT

Climatic environmental conditions, among others, can have an influence upon the reliability of equipment and components in electronic telephone exchanges. This relationship has been statistically studied using vast amounts of data gathered on a national level. A meteorological data base concerning 40 sites equally distributed on French territory, has been obtained from the National Meteorological Office. The information recorded is : the temperature, the relative humidity and the atmospheric pressure which have been recorded every three hours during 1986 and 1987.

In order to study the correlation between meteorological data and reliability, the telephone exchanges have been classified in 9 geographic zones, corresponding to 9 stock yards for board replacements to which the failed equipment is sent before going for repair. Statistical processing has been carried out several times, more particularly, by factor analysis and automatic classification. All of this underlines the correlation which exists between absolute humidity and reliability (Figure 7).

Absolute humidity and reliability of principal boards according to 9 geographic zones (for one electronic telephone exchange).

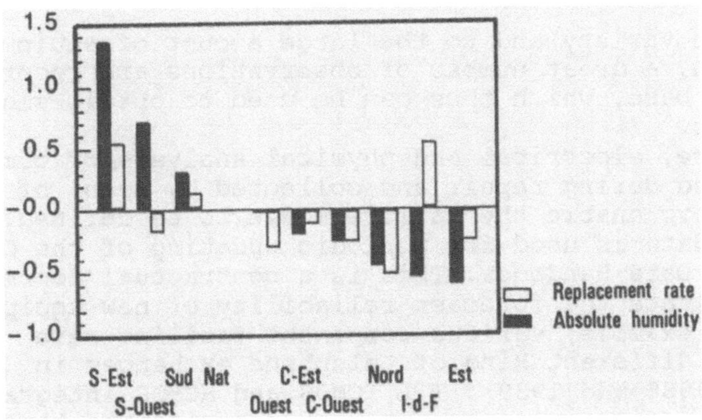

Fig. 7 - Deviation from the national average of a gramme of water per cubic metre of humid air for humidity and of the number of boards replaced each month for reliability.

Humidity induces the corrosion of electronic components. This phenomenon is accentuated by pollution and dust which explains the relatively poor reliability of certain telephone exchanges, in particular, in the Ile de France area.

The harmful influence of a high relative humidity upon the reliability of the boards has been high lighted. However, these tests have shown that the phenomenon of electrostatic discharges caused by low hygrometry does not generally result in a noticeable decrease in the reliability of equipment for the type of telephone exchanges studied.

5.4. QUALITY OF SUPPLY CONTRACTS

During the last few years, the PTT Administration has included guarantee clauses in supply contracts.

For example, these clauses stipulate that, if design or systematic faults occur in more than 10 % of the switching equipment during the guarantee period, the supplier must modify the entire batch at his own expense.

5.5. COMPONENT RELIABILITY

The field failure analysis system is used to determine the reliability of components from the information on repairs, the description of nomenclatures and the population of each type of equipment. Reliability figures can be set out for a particular model or for a set of components having the same technology.

Due to the variety and to the large amount of equipment in operation, a great number of observations are recorded in the data base, which thus can be used to obtain significant results.

Furthermore, electrical and physical analysis of components changed during repair and collected by means of the rejection form enable the failure modes to be defined.

All this data is used for periodic updating of the CNET's Reliability Data Handbook. This is a contractual document used to evaluate the forecast reliability of new equipment.

By way of example, various component families were observed in different kind of telephone exchanges in France, in 1988 and 1989 : TTL, CMOS and HCMOS integrated circuits, memories, microcomputers and associated terminals, transistors, diodes, photocouplers, passive components. Observing numerous devices in the field enabled significant results to be obtained. As can be seen from fig. 8 and fig. 9, the values predicted for diodes, transistors, bipolar integrated circuits, memories, and most of the capacitors used in telephone exchanges were underestimated.

The replacement rate for inductors, aluminium electrolytic capacitors and miniature relays exceeds the values predicted. As a matter of fact, numerous simultaneous replacements of relays are made.

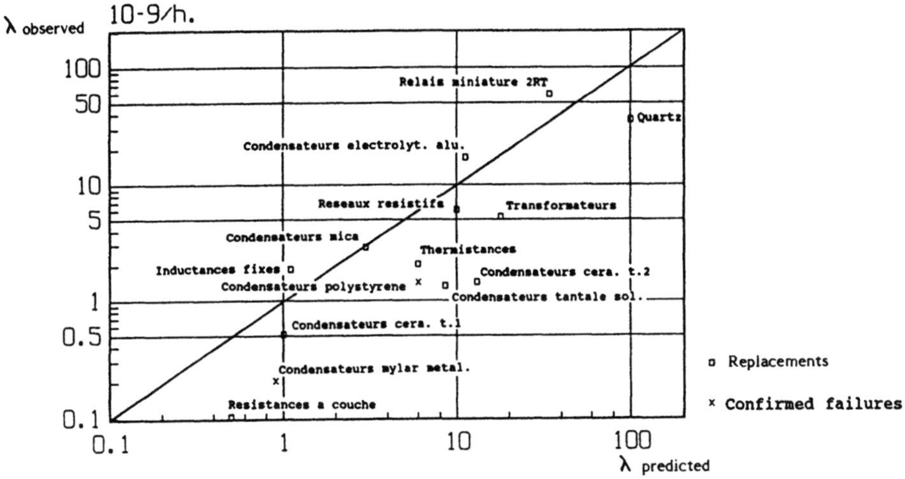

Fig. 8 - Comparison between predicted reliability and observed reliability in 1988 (Passive components)

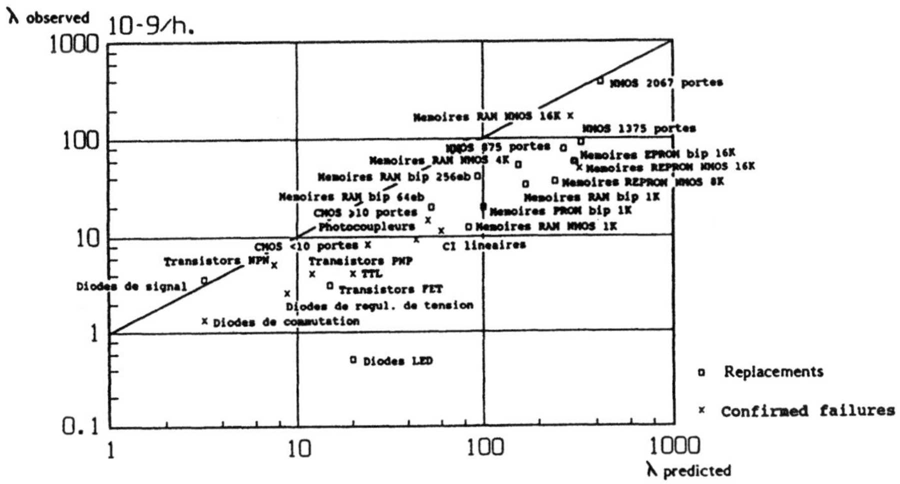

Fig. 9 : Comparison between predicted reliability and observed reliability in 1988 (Active components)

The constant improvement of the reliability of electronic components and also the emergence of new technologies make it necessary to regularly update the Reliability Data Books.

6. CONSULTING THE COMPUTER-CONTROLLED DATA BASE SYSTEM

The data provided by a computer-controlled system are only fully used when they can be selected by the departments concerned by the application themselves.

This is why the data base (SADE-1G) can be accessed in conversational mode by means of a computer terminal connected via the standard switched network. Up to 12 terminals can be connected at the same time (figure 10).

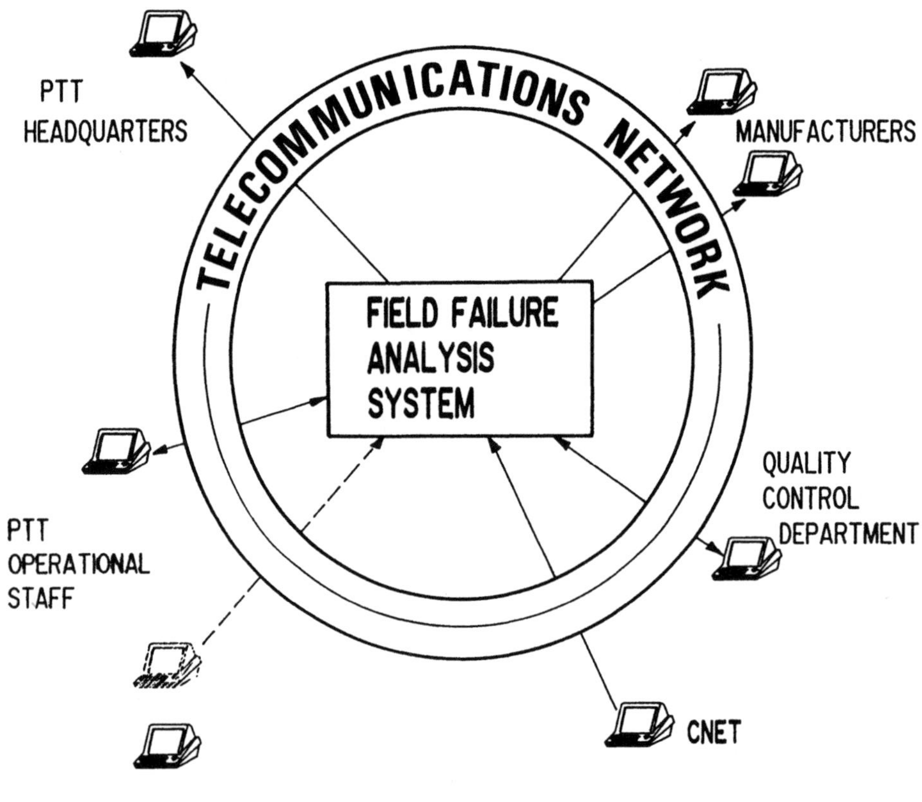

Fig. 10 - Telematic network in reliability and maintainability fields

Users, PTT Administration and manufacturers are offered a wide range of programs chosen according to stated requirements and their running speed.

7. NEW EQUIPMENT AND COMPONENTS

For optical links and optoelectronic components, a joint research programme conducted by CNET (IFE/FAM) and ALCATEL CIT (Cable Transmission Division) was begun in 1985. The purpose is to assess the behaviour of various FRANCE TELECOM 34 Mbit/s multimode optical links (0,85 and 1,3 μm) versus time, to evaluate the operational reliability and the different types of optoelectronic components used.

Traceability on the entire product life cycle was performed using a microcomputer data base. Data was recorded from the beginning (january 1985) until now.

Analysis of the results obtained has taken into account the influence of storage and the failure analysis carried out on rejected components (figure 11).

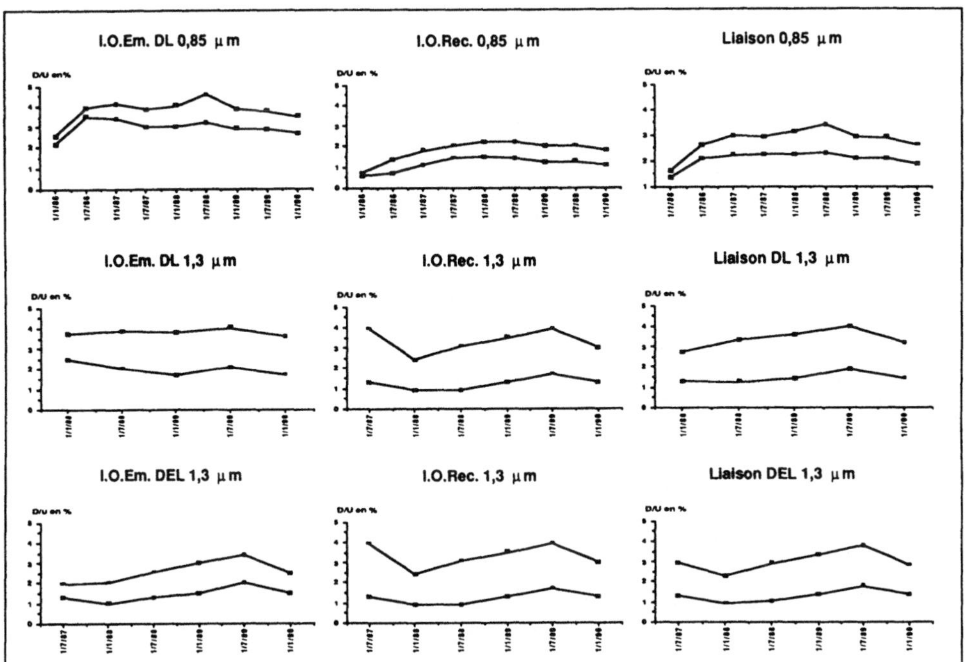

Résultats à la mise en service

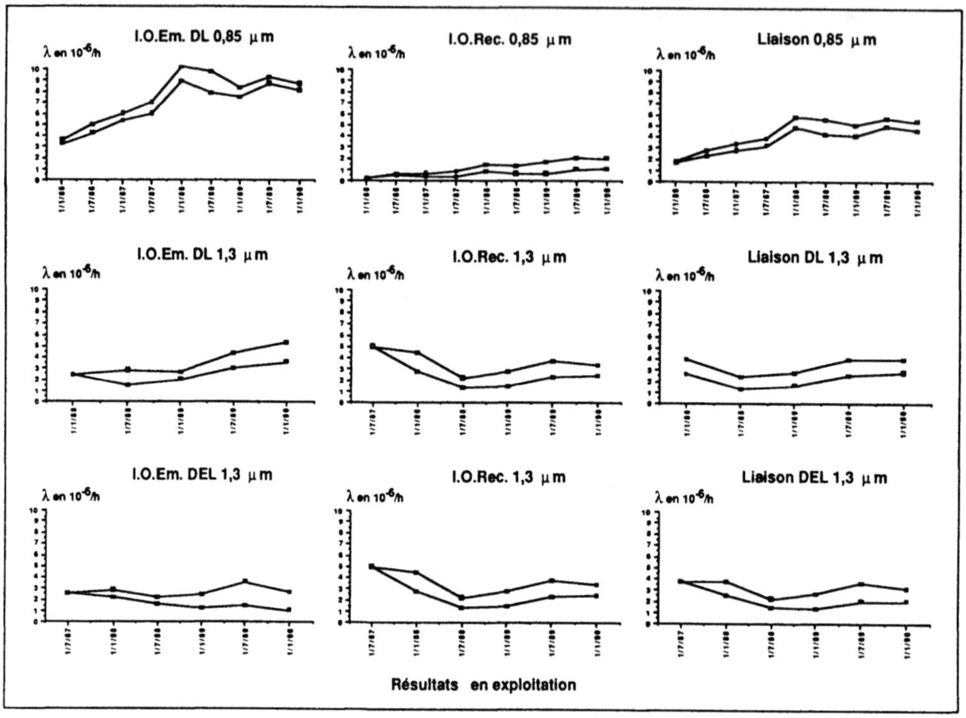

Fig. 11 - Operational results

The results obtained on 0.85 μm optical transmit modules have been used to elaborate a modelling procedure for reliability predictions.

The variation study of the instantaneous λ using the actuarial method was only carried out on the optical modules (0.85 μm) for which sufficient elementary data has been gathered (figure 12).

$$\lambda = \frac{n_i}{N_i\ (T_{i+1} - T_i)}$$

n_i = number of failed optical modules

N_i = number of optical modules which have been operating during the period considered

$T_{i+1} - T_i$ = time interval (two months)

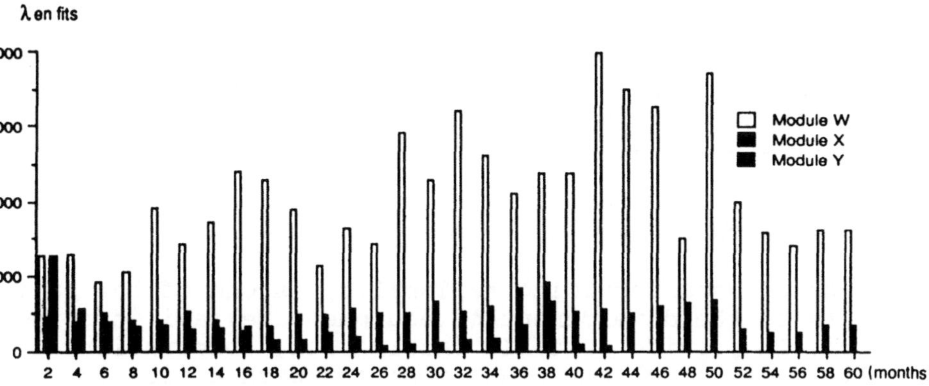

Fig. 12 - Comparison of instantaneous failure rate

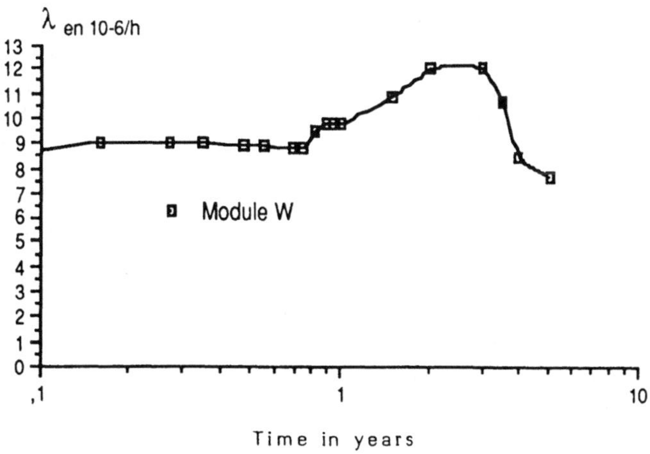

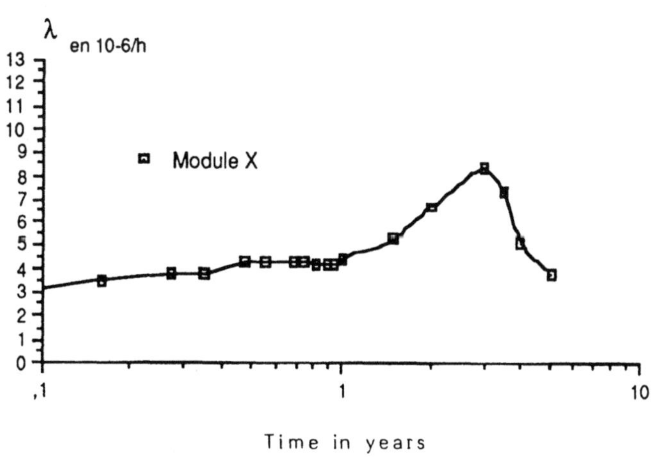

Fig. 13 - Development of instantaneous failure rate

The recording of the evolution of the λ of 0.85 μm optical modules (figure 13) resulted in the combining of two laws : the exponential law and the log-normal law. That is to say, a λ which is almost constant during a period of 18 months and a varying of this λ over 30 months followed by a return to the constant λ .

8. CONCLUSION

The lack of data regarding reliability is one of the major barriers to the development of equipment reliability prediction. To provide suitable information, observation results must be collected and analysed with a view to infer general rules enabling to forecast the future with the best probability. The availability of data is a means of knowing the past events. Yet, the fluctuations in manufacturing, the improvement or drastic changes in the technological development, but also the utilization of new components do not permit to be certain (even with a good probability)of what the future situation will be. Claiming that the reliability data banks will provide a solution to the problem, would be a utopia, but one can reasonably anticipate that an improved knowledge of results obtained, their analysis with a view to knowing the origin of failures, will evidence the interrelation between reliability characteristics and conditions of operation. Reliability cannot be expressed in terms of accurate figures but it will still be possible to improve reliability on the basis of data such as those referred to above. Still, the setting up of reliability data bank is a well-mastered procedure and is of undeniable interest.

THE COMPONENT EVENT DATA BANK

A Tool for Collecting and Organizing Information on NPPs Component Behaviour

S. BALESTRERI
COMMISSION OF THE EUROPEAN COMMUNITIES
Institute of Systems Engineering and Informatics - SER Division
Joint Research Centre
Ispra - Italy

Summary

The Component Event Data Bank - CEDB - is an organized collection of failure/repair events for nuclear and conventional power plants components, in which are specified the "pedigree", operational duties and environmental conditions of the components concerned. The main scope of the CEDB within the European Reliability Data System and its present status are presented.

1. Introduction

An organized collection and exploitation of operating records in nuclear power plants not only is an important tool for reducing the uncertainty ranges in probability estimation for risk assessment studies, but also offers important feedback from operating experience to plant management, architect engineers and manufactures.
At the end of the 1970s the European Reliability Data System project - ERDS - was launched with the aim of collecting and organizing information on:
- operational data;
- reliability data.

The first item, operational data, concerns the continuous collection and organization of events in nuclear plants relevant to safety and availability, i.e.:
- repair actions;
- abnormal occurrences;
- changes in power production.

The second item, reliability data, aims to make available generic failure and repair rates of families of similar components: this data is needed in the field of reactor safety for the estimation of the probability of failure of complex, redundant systems.
Within the ERDS, the Component Event Data Bank (CEDB) is an organized collection of

J. Flamm and T. Luisi (eds.), Reliability Data Collection and Analysis, 125–144.

events such as failures, repairs and maintenance actions affecting major nuclear power plant components. The engineering characteristics of these components are detailed in the CEDB, together with their operational duties and environmental conditions.

The CEDB has been designed to harmonize and centralize the information made available from:

- existing National Data Bank or Utility Data Systems for reliability, maintainability and plant management purposes;
- "ad hoc" data collection campaigns at those plants in which a comprehensive data acquisition system is not yet in operation.

Data can be supplied to the CEDB either in the original system structure and coding of a national data bank (in which case it needs to be transcoded before being loaded into the CEDB database), or directly in the CEDB system structure and coding.

By putting together the operational experience of European nuclear power plants, it was intended to create a set of raw data which can be processed in various qualitative and quantitative ways according to the needs of a user.

The CEDB is now fully operative. At present access to the data is limited to the Data Supplier Organizations. Access for other users is envisaged, subject to special rules designed to preserve the confidentiality of some information.

2. The CEDB (1)

2.1 The Engineering Structure of the CEDB

Since the CEDB channels information coming from various reactor types and from different sources, it was necessary to start by setting up a way to identify equivalent pieces of information.

A set of Reference Classification was established:

- The "System Reference Classification".
  This is a functionally oriented identification of the systems relevant to safety and normal operation of a plant. For each system a unique description of its functions, boundaries and main interfaces with other systems is given. The Reference Classification for LWRs is available (about 180 systems are described and coded); those for Pressurized Heavy Water Reactors and for Gas Cooled Reactors are in draft form.
- The "Component Family Reference Classification".
  This classification groups into homogeneous families components of similar engineering characteristics. About 40 component families have been defined, and up to 20 different engineering attributes have been selected and coded in order to describe the pedigree of each component.
- The "Failure Reference Classification".
  This classification describes a failure event by means of several attributes. Each attribute is coded.

The information reported consist of three main part:
- identification of the "item observed", through its engineering pedigree and installation data (see Fig. 1 and 2);
- identification of the event, through the failure data (see Fig. 3 and 4);
- identification of the operating history of the item, through its operating hours and cycles (see Fig. 5).

The first type of information is reported when the component starts to be monitored (i.e. engineering/pedigree information is provided only once during the life of the component). The second one is reported each time an event/failure occurs on a monitored component. The third type of information is reported, for all the components monitored, on an annual basis.

The physical identification of each plant item is obtained by employing the codings adopted by the utility operating the plant: these codes provide an identification of the position and functions of the component within the plant.

The same code may be used for different items installed in the same position and required to perform the same function(s) as a consequence of successive replacements. These components will be automatically distinguished from each other by their sequence numbers ("progressive number").

2.2 The Informatics Structure of the CEDB

The CEDB informatics structure is designed to take into account the type of data and the Classification adopted. The CEDB, as well as the whole ERDS, has been developed using the Data Base Management System ADABAS (Adaptable Data Base System) of Software A.G.

The original software system was designed and implemented to facilitate user inquiry and analysis of the CEDB. The main computer is an Amdahl 5890-300E running under OS/MVS-XA. It is connected to an internal TP network. The CEDB can be accessed via the public telecommunication netword ITAPAC.

The main programming language used for CEDB is ANSI COBOL. Special informatics tools have been designed and implemented to improve the functionality of the CEDB and to facilitate end-user operations and inquiries.

The large number of tables needed by the Reference Classification adopted for the CEDB has required the setting up of a generalized procedure capable of handling, maintaining, and reporting on a "table data base", where any type of table could be inserted, updated, printed, displayed and protected in on-line and batch-mode.

2.2.1 The Architecture of the Data Bank Figs. 6 and 7 present the CEDB relational and physical/logical structures. The relational structure is based on component information (Fig. 6) and a hierarchical relationship where the component is the root, and operational conditions and failures are dependent levels. Multiple relationships are maintained with other files which can be shared with other ERDS features.

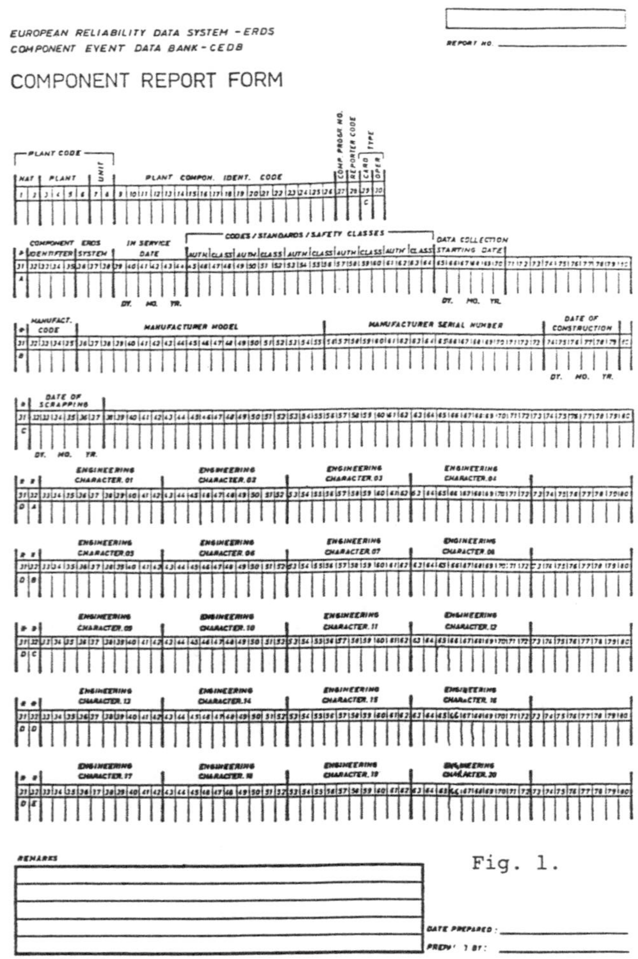
EUROPEAN RELIABILITY DATA SYSTEM - ERDS
COMPONENT EVENT DATA BANK - CEDB
COMPONENT REPORT FORM
REPORT NO.
PLANT CODE
NAT. PLANT UNIT
PLANT COMPON. IDENT. CODE
COMP. PROGR. NO.
REPORTER CODE
CARD OPER TYPE
COMPONENT IDENTIFIER
ERDS SYSTEM
IN SERVICE DATE
CODES / STANDARDS / SAFETY CLASSES
AUTH CLASS
DATA COLLECTION STARTING DATE
DT. MO. YR.
MANUFACT. CODE
MANUFACTURER MODEL
MANUFACTURER SERIAL NUMBER
DATE OF CONSTRUCTION
DATE OF SCRAPPING
ENGINEERING CHARACTER. 01
ENGINEERING CHARACTER. 02
ENGINEERING CHARACTER. 03
ENGINEERING CHARACTER. 04
REMARKS
DATE PREPARED:
PREPARED BY:
FORM CEDB 1 - 1987

Fig. 1.

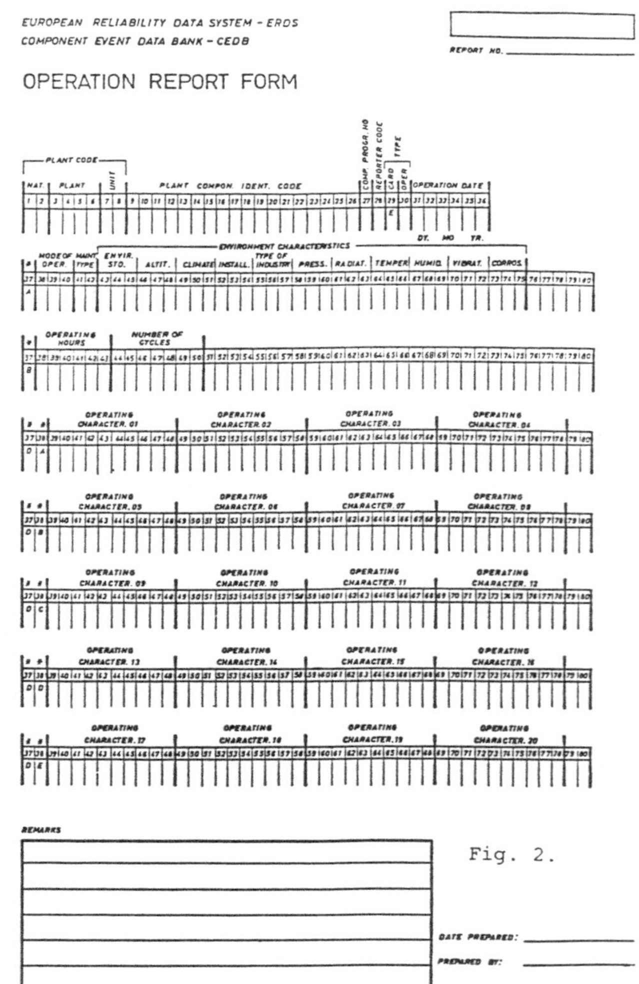
EUROPEAN RELIABILITY DATA SYSTEM - ERDS
COMPONENT EVENT DATA BANK - CEDB
OPERATION REPORT FORM
REPORT NO.
PLANT CODE
NAT. PLANT UNIT
PLANT COMPON. IDENT. CODE
COMP. PROGR. NO
REPORTER CODE
CARD OPER TYPE
OPERATION DATE
DT. MO YR.
MODE OF OPER.
MAINT. TYPE
ENVIR. STD.
ENVIRONMENT CHARACTERISTICS
ALTIT.
CLIMATE
INSTALL.
TYPE OF INDUSTRY
PRESS.
RADIAT.
TEMPER.
HUMID.
VIBRAT.
CORROS.
OPERATING HOURS
NUMBER OF CYCLES
OPERATING CHARACTER. 01
OPERATING CHARACTER. 02
OPERATING CHARACTER. 03
OPERATING CHARACTER. 04
REMARKS
DATE PREPARED:
PREPARED BY:

Fig. 2.

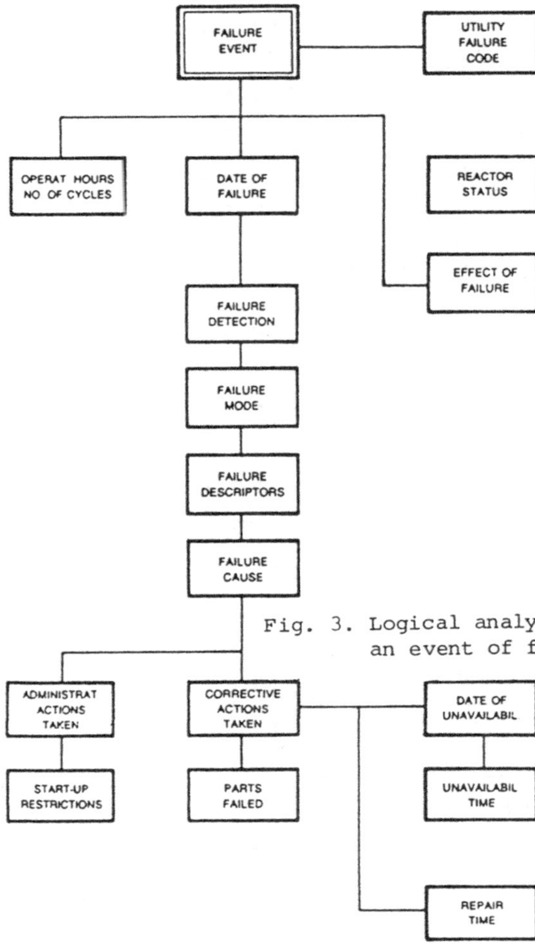

Fig. 3. Logical analysis of an event of failure

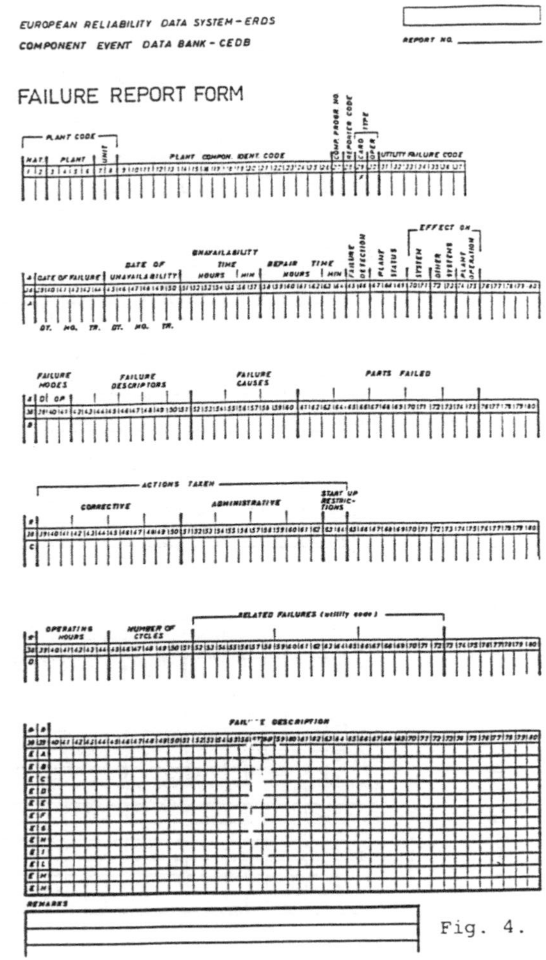

Fig. 4.

EUROPEAN RELIABILITY DATA SYSTEM - ERDS
COMPONENT EVENT DATA BANK - CEDB

REPORT NO.

ANNUAL OPERATING REPORT FORM

PLANT CODE NAT. (1–2)	PLANT CODE PLANT (3–6)	PLANT CODE UNIT (7–8)	PLANT COMPON. IDENT. CODE (9–26)	COMP. PROGR. NO. (27)	REPORTER CODE (28)	CARD TYPE (29)	CAMP. BEG. IND. (30)	YEAR (31–34)	OPERATING HOURS (35–40)	NUMBER OF CYCLES (41–47)
						H		1 9		
						H		1 9		
						H		1 9		
						H		1 9		
						H		1 9		
						H		1 9		
						H		1 9		
						H		1 9		
						H		1 9		
						H		1 9		
						H		1 9		
						H		1 9		
						H		1 9		
						H		1 9		
						H		1 9		
						H		1 9		
						H		1 9		
						H		1 9		
						H		1 9		
						H		1 9		
						H		1 9		
						H		1 9		
						H		1 9		
						H		1 9		
						H		1 9		
						H		1 9		
						H		1 9		
						H		1 9		
						H		1 9		
						H		1 9		
						H		1 9		
						H		1 9		
						H		1 9		
						H		1 9		
						H		1 9		
						H		1 9		
						H		1 9		
						H		1 9		
						H		1 9		
						H		1 9		
						H		1 9		
						H		1 9		
						H		1 9		
						H		1 9		
						H		1 9		

REMARKS

Fig. 5.

DATE PREPARED: ____________

PREPARED BY: ____________

FORM CEDB 4 – 1987

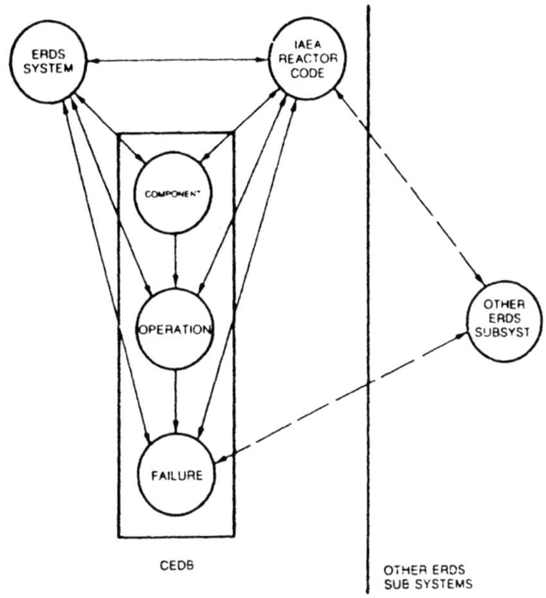

Fig. 6 - Relational structure of the CEDB.

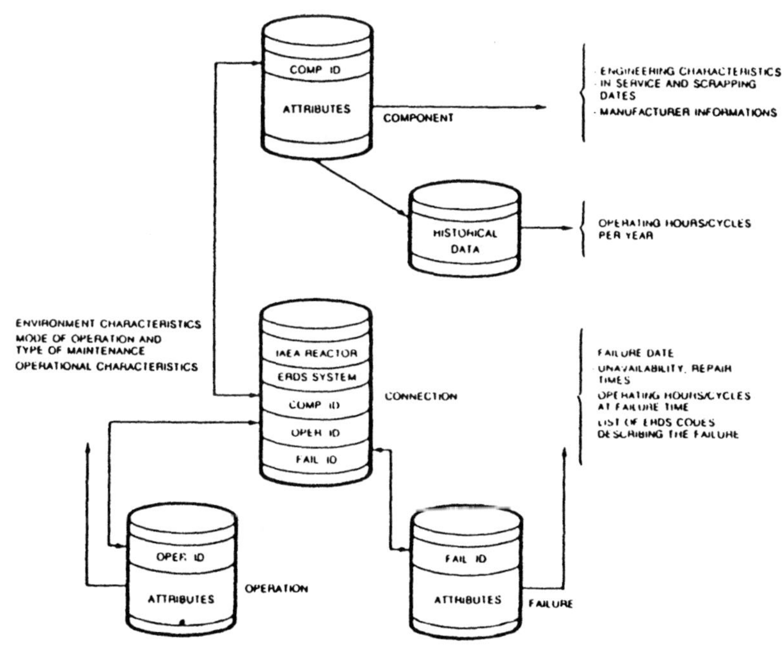

Fig. 7 - Physical-logical structure of the CEDB.

The information content of the CEDB is characterized by three main logical entities, subdivided into four ADABAS files:

- component pedigree data;
- component operation and environment data;
- component failure data;
- a connection file, providing the connection and relations between the data stored in the above-mentioned files.

The "connection file" (Fig. 7) uses internal identification codes (identifiers) to link the information netword. These identifiers provide a unique identification of each component, operation and failure in the CEDB.

2.2.2 Input Procedures An automatic procedure has been implemented for the updating and validation of CEDB data. The set-up controls can be classified as follows:

- formal controls, such as presence of mandatory data, agreement with coding in the reference tables, etc.;
- coherence controls, such as checking of data credibility, correctness of temporal sequences, consistency of data, etc.

Various actions are taken, depending on the severity of the error found. Lists for the correction of the errors detected and for the monitoring of CEDB status have been set up. The updating procedure consists of modular programs, set up by "top-down" structural programming techniques.

2.2.3 Output Procedures On-line and batch access to CEDB is available to foreign contributors/users through the ITAPAC Network. Owing to the complexity of the CEDB structure and content, and in order to avoid problems or misinterpretation by the users, an original query and analysis system has been designed and implemented. The system enables the selection of any information sample desired and the calculation of the required reliability parameters. Qualitative analysis of the data selected can also be performed. Passwords and privacy protection devices have been set up within the system.

2.3 Interrogating the CEDB (2)

The mechanism for handling CEDB enquiries has been designed to facilitate the commonest type of enquiry, namely those aimed at computing statistical parameters of reliability. It therefore follows the hierarchical relationship: component-operation-failure.

The concept of "selection" has been introduced as the elementary logical unit of enquiry (Fig. 8). A selection is a three-step search using criteria on components, operations and failures, producing three sets of components and a set of failures.

The 1st set of components is the result of stepwise refinements on the components, the 2nd set is a subset of the first one obtained by stepwise refinements on the operation and the last one is a subset of the 2nd one obtained by stepwise refinements on the failures. At this

lowest level the set of failures of the selection is also available.
In a session one can refer to any previous selection which is currently available. In re-examining the selection one can refer to either of the two sets of components or to the set of failures (with their related components) and use this set as the starting point for further enquiries. It is important to note that the various statistical computations are done on the last set of components selected during the session, before the selection of the failure events.
Help facilities are provided at any moment during the session, this allows, for example, the consultation of codification tables, engineering characteristics classifications, or the history of previous selections. In Fig. 9 the succession of the stages of a full enquiry session is shown. In Fig. 10 the commands which can be used during the session are listed.

2.4 The Statistical Processing of the Data (2)

The CEDB has a suite of programs for on-line and batch processing of the data. These programs can be called by the command STAT (CLE, BAE, ENT, TST) during any enquiry sessions.
The statistical processing includes at present:

a. Point and interval estimation (for complete and censored samples), of:
   - a.1 Constant reliability parameters (time-independent failure rate in operation, constant unavailability = constant failure rate on demand, time-independent repair rate).

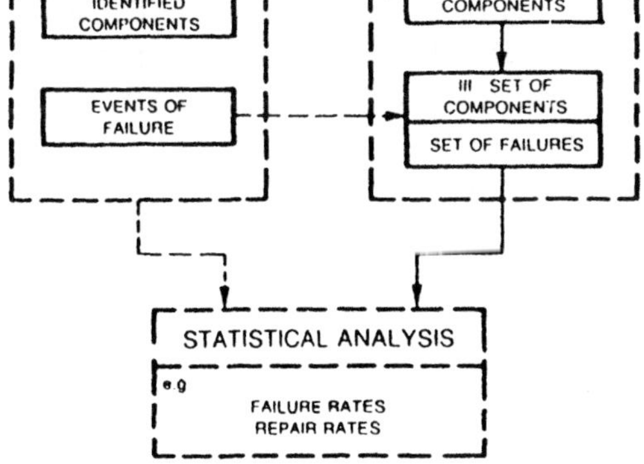

Fig. 8 - General logic of the inquiry.

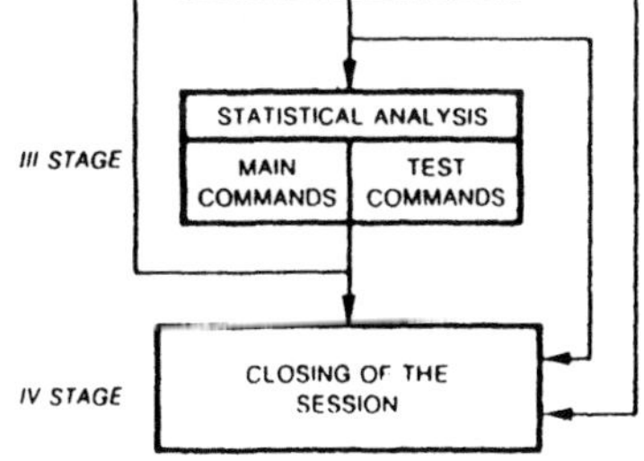

Fig. 9 - Inquiry session.

I° STAGE	II STAGE		III STAGE		IV STAGE
OPENING OF THE SESSION AND GENERAL INFORMATION COMMANDS	SELECTIONS		STATISTICAL ANALYSIS		CLOSING OF THE SESSION
	MAIN COMMANDS	OTHER COMMANDS	MAIN COMMANDS	TEST COMMANDS	
LOGON					
HELP DISPLAY					
	SELECT with criteria OR Snn END SELECTION	CANCEL DELETE DISPLAY DISTRIBUTION HELP SHOW DISTRIBUT + SHOW			
			STAT CLE STAT BAE	STAT ENT TST SAS	
					LOGOFF

Fig. 10 - Session inquiry commands.

- Bayesian parametric approach (with priors: beta, uniform, loguniform, lognormal, histogram);
- classical approach (maximum likehood, confidence interval).

a.2 Non-constant reliability parameters (time-dependent failure rate in operation, non constant failure rate on demand, time-dependent repair rate) by the Bayesian non parametric approach (with the prior identified by a sample of times to failure or by a failure-time analytical distribution).

b. Test of hypothesis on the law of failure and repair time distribution:
   - exponential (for complete and censored samples);
   - Weibull, lognormal, and gamma distribution, increasing failure rate, decreasing failure rate (only for complete samples).

Effective graphical tools can give on-line the representation of an observed time-dependent failure rate; of the prior and the posterior distributions (Bayesian parametric approach); of the cumulative failure distribution function F of the observed sample, the prior and the posterior sample (Bayesian non-parametric approach), etc.

In refining a selected sample of failures for a statistical analysis, the analyst can retrieve and review each event, to identify, and possibly delete from the sample, those failures which appear not to be independent.

3. Source of Data and Status of the CEDB September 1990

The data suppliers to the CEDB are listed in Tab. 1:

Tab. 1 - Data Suppliers to the CEDB.

Data Supplier	Plant
SRDF (EDF-)	NPPs Fessenheim 1-2 Bugey 2-3-4-5
ENEL (I)	NPP Caorso CONV. La Casella 1-2-3-4
KEMA (NL)	NPPs Dodewaard Borssele
ATV (Vattenfall-S)	NPP Ringhals 2
CEGB (UK)	NPP Odbury A-B
EBES (B)	NPP Doel 3
UNESA-E	10 plants

Note : In the conventional power plant of La Casella (I) data is being collected on components of the feed water system only, due to their similarity with components operating in analogous systems in Nuclear Power Plants.

The Status of the CEDB at September 1990 is the following:

- Total number of Components 6517
- Total number of Events 5868
- Operating experience:
  - . Average calendar years 6
  - . Reactor-year 108
  - . Component-years 38953
- Component families (with the number of items loaded into the CEDB in brackets). Electromechanical Actuator (638), Hydraulic Actuator (131), Pneumatic Actuator (514), Battery (35), Blower (3), Circuit Breaker (258), Clutch (30), Compressor (26), Electri Motor (511), Internal Combustion Engine (8), Heat Exchanger (138), Filter (41), Electrical Generator (30), Motor Pump Unit (4), Pump (521), Rectifier (27), Safety valve (349), Steam Generator (20), Switchgear (29), Accumulator/Tank (98), Transformer (20), Turbine (25), Valve (3060).

The evolution of the contents of the CEDB during the last 4 years is shown in Tab. 2. In the last column of the same table is given the status of the Data Bank predicted for the end of 1990.

Tab. 2 - Evolution of the content of the CEDB.

	End 1985	End 1986	End 1987	End 1988	Oct. 1989	End 1990
Total # of components	4580	4670	5180	6020	6515	13500
Total # of failures	3225	3825	4290	4720	5825	7500
# of plants	11	12	16	18	18	30
. average calendar years			4.9	5.2	6.0	
. reactor-year			78	94	108	
. component-year			25400	31300	38750	

The significant expansion of the CEDB expected from 1990 is mainly due to the data being collected in 10 Spanish NPPs under the supervision of UNESA, the Association of the private Spanish Utilities and in 4 conventional power plants in the U.K. by the former CEGB.

4. Management of the CEDB

The CEDB is managed by the JRC, Ispra (which employs a staff of engineers, informaticians and statisticians of the Institute of Systems Engineering and of Informatics).
In October 86 a Steering Committee (S.C.) of the CEDB Data Suppliers was formed under the chairmanship of EDF. At present full members of the S.C. are ATV-VATTENFALL (S), CEGB (UK), ENEL (I), JRC, KEMA (NL), SRDF-EDF (F) and UNESA (E). UNESA (E) holds the Chairmanship of the S.C. up to the end of 1990 (3).

5. Implementation of Special "Tools" for Data Collection and Management

As described in Chapters 1 and 3, the possible sources of data are existing National Data Banks or Utility Data Systems, and "ad hoc" data collection campaigns. In either case some technical problems should be taken into consideration.

5.1 Existing National Data Banks and/or Utilities Data Systems

The information concerning the component characteristics, its environment/installation, its operation duties and related failure events are already collected and organized according to the aims, structure, classifications and application of the source Data Bank. All this data has to be adapted by transcoding to the structure and classifications set up for the CEDB. To accomplish this task, an in-depth knowledge is needed of the structure of the original source reports, of the philosophy and the approach used in the reference classification, and of the procedures according to which data are reported by the Utility.
Problems may arise when the information collected by national data systems is not sufficient for exhaustive compilation of CEDB report forms.
Additional information may be requested, concerning mainly environmental conditions, records from time/demand counters, maintenance planning programs, test frequency of stand-by systems, and so on.
Two examples:

1. Data supplied by the French Système de Recueil de Données de Fiabilité SRDF-EDF (6 NPPs, about 3000 component items and 2700 events) were transcoded manually. The transcoding rules from SRDF to CEDB had already been defined as a result of a previous study. Some other data, mainly operating data, were added later.
2. Data supplied by the Swedish ATV-Vattenfall (1 NPP, about 600 component) were already in CEDB formats. The CEDB codings were obtained by means of a simple ad-hoc transcoding program designed by Vattenfall. Integration of new data from the plant was performed manually.

5.2 The JRC Semi-Automatic Transcoding System (4)

The transcoding of huge amounts of data is a very time-consuming task, which requires a great deal of expertise. In fact the logical entities in the various databases do not necessarily correspond because of the different design philosophies. Consequently, the transcoding work does not simply consist of finding a correspondence between codes, but in representing in one system a logical entity which is originally described in another system.
In 1986 a feasibility study for the development of a generalized semi-automatic transcoding system to be applied to the European Reliability Data System was started. The principal aim of such a system would be to reduce the amount of manual transcoding needed. It was also expected that the quality of the transcoded data would be improved.
The study addressed the construction of a generic system, independent of any specific application - i.e. a particular source database and a particular target database within ERDS. The intention was to use expert systems techniques. After a favourable conclusion from the feasibility study, the realisation of such a system was started; it will be operative in 1990. The first application will be using the Component Event Data Bank as ERDS target

database and the French Système de Recueil des Données de Fiabilité, SRDF-EDF as source database.

The main characteristics of the system are the following:

- It deals with transformations of format (measurement units, dates), and of contents, considering not only 1-1 relations between codes, but also more complex relations, represented by conditional expressions and relations involving algorithms of transformation.
- It works in a semi-automatic way. When an automatic transformation of data is not possible, the system displays relevant information and possible choices to the user; the choice made by the user is subsequently checked by the system.
- It gives a trace of its reasoning; i.e. the sequence of rules it has applied.
- It provides a user-friendly procedure for the transcoding rules, intended for a non-programmer expert.
- It produces input forms for the target database and makes no validation checks on the data transcoded (these data must be checked on passing through the target database input procedure). The general scheme of data flow from the national database towards the Component Event Data Bank (CEDB), is shown in Fig. 11.

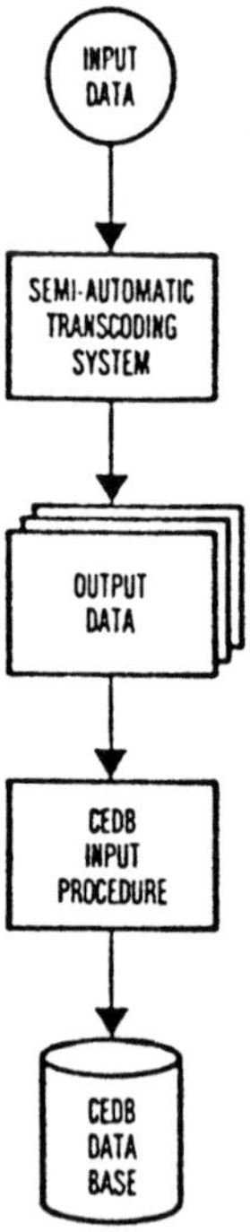

Fig. 11 - Data flow from a national database to the CEDB.

The basic software system consists of a set of programs which are applicable independently, namely:

- the transcoding program, running both in batch and on-line mode, which can be considered as a kind of inference engine for the system in the sense that it makes control decisions on the use of the knowledge base (Fig. 12);
- the two procedures used to define a rule for input, i.e.:
  . the input and output model definition procedure,
  . the rule definition procedure which translates the rule language into a suitable internal representation.

The function of these two procedures is similar to the knowledge acquisition module of typical knowledge system architectures.

The languages used for the implementation of this basic software are NATURAL (Software AG) and COBOL.

To recapitulate, the knowledge of the system consists of models and rules. These are represented as data structures stored in a database under ADABAS with the help of the above definition procedures, and are in turn managed by the transcoding program.

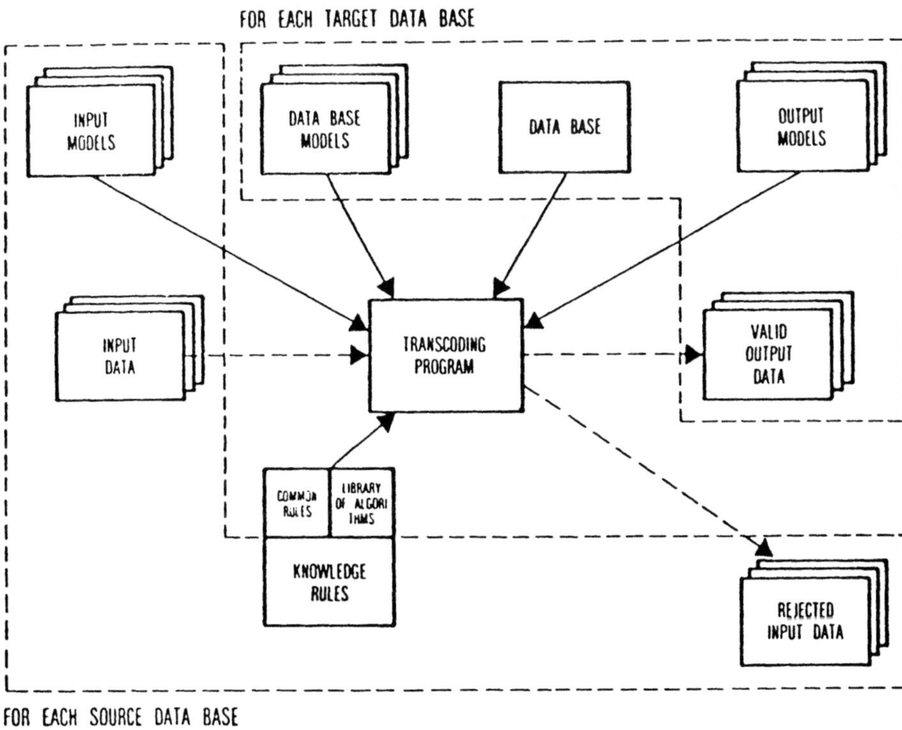

Fig. 12 - Overall transcoding system architecture.

5.3 Ad-Hoc Data Collection Campaigns

Data collection campaigns can be a useful source of good quality information, provided that:

- a reliable system of archives exists for component data information retrieval;
- failure data and/or order sheets are accurately recorded and categorized;
- the on-site personnel have a good knowledge of the CEDB.

CEDB staff provide detailed training of the personnel involved in the data collection, together with some help for identifying the sources of data in the archives. In some cases a redesign of local work order sheets has been suggested, in order to have a more complete and exhaustive record of important pieces of information useful not only for the CEDB but also for plant management and maintenance policy. Special "Guides" have been published to help the Reporter in compiling the Report Forms (e.g. for pumps and valves) (5) (6). Other guides are under preparation (in particular for the compilation of the failure report form).

6. Examples of Collaboration between the CEDB and National Organizations

6.1 CEDB/ENEA-VEL (I)

During the period 1985-1988 the department for fast reactors in Bologna of the Italian ENEA (Ente Nazionale Energie Alternative) adopted the CEDB structure for setting up their ENEA Data Bank for Fast Breeder Reactors (7).

This Data Bank is used as a tool for supporting several activities in the field of fast reactors, such as:

- development of reliability studies;
- improvements in design and maintenance policies for special components;
- feedback of experience from experimental reactors into the design of power reactors.

Some interesting modifications have been introduced in the design of the Data Bank. They can be summarized as follows:

- extension of the engineering characteristics;
- addition of new types of components;
- the possibility of following the component during its life through its movement around in the plant;
- more component maintenance information, recording the parts that have been changed and the reason for the substitution.

As a result of the collaboration between CEDB and ENEA these improved engineering classifications can now be adopted in other European Data Banks.

6.2 CEDB/UNESA (E)

In July 1987 the Spanish Nuclear Safety Council approved the proposal of UNESA (the Association of Spanish Electrical Utilities) to store in the CEDB the data on certain components relevant to safety of all Spanish Nuclear Power Plants in operation.
The principal aim of collecting these data was to support the realisation of the PRA for each plant, required by the Safety Authority.
The enlargement of this activity to other systems and components relevant to plant availability and maintainability can be envisaged as a second step of this collaboration.
The design of the procedures for collecting data in the plants has been carried out by TECNATOM (an engineering company owned by UNESA), and fully agreed with the CEDB.
The data are centralized by TECNATOM and after some technical controls (by means of a dedicated software package), they are sent of magnetic support to the JRC at Ispra.
A rough estimate of the volume of data to be supplied is of the order of 7000 component year per year. Data on failures will be collected as from the beginning of 1989. No historical data collection is envisaged at present.
The internal procedures for data collection required, in some cases, a foundamental redesign of the work order sheets and a revision of the management of technical information.
All Spanish Utilities will be connected to the CEDB via the public telecommunication network IBERPAC-ITAPAC.

6.3 CEDB/NNSA

In the framework of the "People Republic of China - CEC Energy Cooperation Program" established on October 1987, one of which aims is to create in China a "Nuclear Safety Information Centre" (NSIS), the JRC was asked by DG XVII to collaborate with the Chinese "National Nuclear Safety Administration" in order to set up in China a system of data banks analogous to the ERDS.
Since then, a series of technical meetings have been held, in Ispra and Beijing, to offer engineering advice to the NNSA to facilitate the formation of data banks of the type "Abnormal Occurrences Reporting System" (AORS) and "Component Event Data Bank" in the ambit of NSIS. In-depth training was also given to some Chinese engineers who visited the JRC.
It has also been decided to supply to the NNSA a version of the AORS and CEDB software, ready to be installed on the CYBER computer at the headquarters of the NNSA in Beijing.
Because of the basic differences between the NNSA computer (a CYBER 180/830) and the JRC computer - which run under different Operating Systems - and between the Data Base Management Systems installed on these computers (IM/DM of Control Data in

Beijing, and ADABAS of Software AG at Ispra) it was necessary to implement two "ad hoc versions" of the two data bases (software and data structure), in order to create the required compatibility for an exchange of data to be agreed with the CEDB Data Suppliers.
Tab. 3 shows the most significant characteristics of the CEDB versions working at Ispra and in Beijing.

Tab. 3 - CEDB, main features of the ADABAS and IM/DM version.

THE COMPONENT EVENT DATA BANK	at JRC-Ispra	at NNSA-Beijing
Computer	AMDAHL 5890-300E	CONTROL DATA CYBER 180/830
Operating System	MVS/XA	NOS/VE
Data Base Management System	ADABAS	IM/DM (Information Management/ Data Management)
of type	inverted file structure	relational
Input Procedure	batch mode	on-line mode
	4 report forms	4 screens
	SAME FORMAL AND COHERENCE CONTROLS	
Enquiry Procedure	Query and Analysis Language	using DM/FQM (Fundamental Query and Manipulation Language)
	Retrieval of data + Statistics	without Statistics for the moment
of type	user oriented language of commands	general purpose language of IM/DM (SQL-like language)
Programming Languages	mainly Cobol Natural Fortran for Statistics	Fortran DM/FQM

The IM/DM version was successfully installed on the CYBER Computer in spring 1989 by the informaticians of the CEDB staff.
The technical collaboration with the NNSA will certainly continue in future in the framework of the activities of the Commission.

6.4 CEDB/ENEL-CRTN (I)

The collaboration between the CEDB and ENEL-Centro Ricerche Termonucleari (CRTN) started at the beginning of the ERDS project. The first data collection campaign, performed in 1982 in the NPP of Caorso, gave a quite large and exaustive sample of data for the testing of the structure and software of the CEDB.
Since 1985 the ENEL-CRTN performed a series of campaigns in Caorso and gave its supervision to the similar activity performed by the ENEL-Settore Produzione e Trasmissione in the conventional power plant of La Casella (four units, 320 MWe each unit). All these activities have been performed in close collaboration with the JRC Ispra.
The CEDB database is now the system in which the ENEL centralizes its data; an extensive work of analysis of these data is envisaged by a joint team experts.

7. The CEDB as a Source of Reliability Parameters

As mentioned in Chap. 2.4 the CEDB data can be processed by means of the statistical programs implemented in the inquiry procedure.
Another feature is the possibility of preparing a set of data (e.g. all pumps or all pumps of a certain type) in the format accepted by the package SAS "Statistical Analysis System", a powerful tool for data treatment.
A lot of work is being performed in the field of analysis of the operating experience of electrical power plants (8).

8. Conclusions

The experience gained in development and application of CEDB has shown that the data modelling structure and the database software which both support the Data Bank are a suitable and versitile tool for collecting and handling reliability data coming from different sources.
This is confirmed by the many Countries and Institutions which have adopted either the CEDB classification and/or the whole informatic system.
Furthermore the CEDB has proved to be quite successful as a mean for exchanging information on component behaviour in various applications.
The enlargement of the CEDB and its wider use will eventually prove its inherent capability for providing sound component reliability models and parameters.

References

(1) Component Event Data Bank Handbook (1984). JRC T.N. 1.05.C1.84.66.

(2) Balestreri S., Carlesso S. and Jaarsma R. (1988). Inquiry and statistical analysis of the CEDB by means of a terminal. User Manual. JRC T.N. 1.05.C1.85.128.

(3) Besi A. (1987). Rules for CEDB operation. JRC PER 1091/87.

(4) Carlesso S. (JRC), Melagrano B., Montobbio M. and Novelli C. (SOFTECO, Genoa) (1988). Semi-automatic transcoding system applied to ERDS. 19th International Software A.G. Users Conference, Vienna October 1988.

(5) Balestreri S. (1989). CEDB Handbook, Engineering Reference Classification and General Guide. PUMP (Doc. 1615/88).

(6) Balestreri S. (1989). CEDB Handbook, Engineering Reference Classification and General Guide. VALV (Doc. JRC PER 1646/89).

(7) Righini R. (ENEA-VEEL) and Sola P.G. (NIER) (1987). How to use the ENEA Data Bank. Doc. ENEA RT/VEL/87/3.

(8) Besi A. and Kalfsbeek H. (1989). Development of methods for the analysis of NNP operating experience. Reactor Safety Research Seminar, Varese, November 1989.

PREDICTION OF FLOW AVAILABILITY FOR OFFSHORE OIL PRODUCTION PLATFORMS

G.F. Cammack
British Petroleum
Moor Lane
London

Use is made of availability techniques from the early stages through to completion of the design of offshore oil production installations. The techniques are used to predict product flow levels and to compare different equipment configurations. This enables the benefits of the increased production which can be gained from installed spares and multiple flow paths to be weighed against the penalties of capital cost, weight and space, which result from the addition of extra equipment.
This paper will address the methods that one company uses to predict the flow availability of a new design, the difficulties in obtaining useful data and indicate the means used to demonstrate cost effectiveness of different equipment arrangements.

1. Method

The availability analysis can be carried out using the conventional analytical methods used to calculate the availability of repairable systems (see Ref. 1 for basic explanation of method and mathematics). If the plant contains major variables as inter-stage storage, non-regular feed supply or export routes or other non-constant states then the use of Monte Carlo simulation may be necessary.
In either case equipment failure and repair data is required in order to develop a basic model of the system under review.
The studies can beneficially commence at the conceptual design stage of a project and be developed with increasing levels of accuracy and detail as the design develops and equipment lists and choice of type and manufacturers of equipment are finalised. The early studies and models being modified in line with the design changes.
To commence the studies, provisional flow diagrams and equipment lists need to be obtained. At the early stage of design, equipment lists will frequently only show size and type of equipment needed and not numbers of units or specific designs, but nevertheless useful models can still be developed at this time.

J. Flamm and T. Luisi (eds.), Reliability Data Collection and Analysis, 145–159.

It is also necessary to be aware of the intended operating and maintenance philosophies as these have a major influence on repair and response times.

At this stage the receipient of the availability study must confirm the position of the study boundries. The export of oil or gas from a platform is dependent upon not only the process trains and export facilities on the installation itself but also upon the wellflow, the export pipeline and the reception facilities on shore. Nowdays an increasing number of installations export their product via a pipeline and other facilities belonging to a third party, the availability of which is beyond the control of the first party. The target availability of the new installation needs to take acount of down time imposed by third parties.

Long term production variables as drilling programmes, field production profiles and needs for future gas compression or enhanced oil recovery (EOR) etc need to be known. These variables determine the demands on utilities, in particular power requirements which may increase or decrease with time or if there are any peaks of power consumption in the field life.

It is necessary to state at what period of the installations life that the study refers. This will usually be during the period of normal operating conditions after commissioning has been completed and any early equipment failures rectified. At this stage equipment is assumed to fail randomly giving an constant failure rate.

An elementary, but very necessary requirement, is to agree if calculated availabilites do or do not include scheduled shutdowns either of individual items of equipment or of the whole field.

Using this information and the process flow diagram (shown schematically in Fig. 1) an availability flow diagram can be produced (Fig. 2).

The latter will differ substantially from the process flow diagram as it incorporates the supporting utilities that enable the plant to function as well as the equipment that features in the process trains. Thus items such as power generation, waste disposal, flare systems, water winning and handling systems, instrument and control requirements, storage and export routes etc. all have to be taken into account, for without any of these production cannot continue.

The blocks or modules of the availability flow diagram are further divided into more detailed blocks (Fig. 3). These blocks are further refined until individual machines, plant items and duplicity of flow paths are shown (Fig.4). Eventually the availability engineer is satisfied that the individuality of the area under consideration has been identified and the flow availability that the design allows can be quantified.

With reference to the relevant data the probable failure rates, restoration times and time required for scheduled maintenance can all be taken into account and the likely availability of individual components calculated.

Reference to Fig.4 shows certain items given an availability of unity. This can indicate:-

an area eg. well flow, over which the design engineer has no control and will be the same in all the variations in the design.

- a piece of equipment e.g. the tank in Fig.4 whose reliability is such that it can be considered to be 100% available at all times in the field life, except for schedule maintenance.

- an item for which at this stage of the study no data can be found.

In the studies availability is usually shown to four decimal places as it is unrealistic to pretend to greater accuracy (remember 0.0001 of a year is less than one hour). It is necessary however to calculate to at least seven decimal places to reduce rounding errors.

By combining individual component availabilities the flow availability of larger blocks and eventually of the whole platform, can be obtained.

Flow availability is considered as the quantity of oil and gas exported over a given period compared to the theoretical maximum that the plant can process in that time.

Flow availability can be expressed as:-

$$A_{Flow} = \frac{\text{Volume of product exported}}{\text{Theoretical maximum exportable volume}}$$

Flow availability is a function of the plant availability and the plant maximum design throughput. Thus the extra capacity required of a plant to make up production lost due to process unavailability can be evaluated.

Plant availability is governed by equipment availability which is calculated according to the commonly used:-

$$A_E = \frac{\text{Mean time between failures}}{\text{Mean time between failures + mean time to restore}}$$

Mean time to restore is a combination of actual repair time and time to re-commission the plant to full flow at correct specification. To obtain "correct specification" is usually much less onerous with offshore production than it is in, for example, a chemical plant, where restoring the product stream to a critical specification may take much longer than a simple repair or restart of a tripped machine.

Conversely "time to restore" can be much longer offshore than onshore. In an offshore situation a simple repair can be very time consuming if men have to be mobilised and transported, together with materials, by helicopter or ship, from the shore base or a parent platform to the repair site.
Examination of the flow availabilities calculated for the individual blocks will indicate those features of the design which have the major effect on overall flow availability.
These areas can then be addressed and the benefits of improvements compared to the associated costs. These latter include space, weight, manning requirements, etc. as well as monetary costs.
The availabilities are calculated using algorithms based on the classical analytical techniques explained in many text books (Ref. 2,3,4).
At BP the algorithms normally used have been incorporated into a specifically written software program which can be used on a PC. There are however programs which will assist in the studies becoming available commercially.
The use of a suitable program is considered to be essential if these studies are to be completed in a reasonable time frame and if arithmetical errors are to be avoided.

2. Size? Number? Type?

The analysis emphasises the common design problems of deciding "how many", "what size" and "what type" of a particular unit should be selected.
It is a frequent requirement in a design to decide if the optimum combination of machines or vessels at a particular location should be 1 x 100%, 2 x 50%, 3 x 50%, or whatever.
At the simplest it can be assumed that one machine will give a lower likelihood of failure than two machines but the consequences of failure of a 1 x 100% machine will be greater than the failure of one machine from a 2 x 50% or 3 x 33% configuration.
In general, three 33% units will be more expensive than two 50% units which in turn will be more expensive than a single 100% capacity unit. Also three machines will weigh more and take more space than two, etc., which are very important considerations in platform design. The greater number of smaller machines also generates the need for extra pipework, cabling, instruments, etc, which must be taken into account (Ref. 5). An increased number of machines also leads to increases in maintenance requirements and operating costs.
The justification for the choice of a particular combination of machines will be governed principally by availability considerations and comparing the benefits that accrue from the increased production that results from improved availability with the likely cost and other

penalties that go in parallel.

3. Example

The use of the techniques can be illustrated by deciding the optimum number of gas turbine driven generators for a hypothetical platform, which has a maximum power requirement of 14MW. Assume that the favoured power generator combination has an output of 14MW and an installed cost of £13MM. The power requirement can thus be met by one machine. This can be considered as a "zero based" option. It is normally accepted however that more than one machine will be desirable, as otherwise scheduled shutdowns alone will give an unreasonable downtime and associated production loss.

Other combinations and types of machines will therefore need to be considered. A suitable smaller machine is available with a power output of 7MW and an installed cost of £8mm. Uprated versions of this machine will provide 9MW at an installed cost of £8.5mm or 10.5MW at £9mm.

These smaller machines provide 50%, 65% or 75% of the required power, but obviously more than one will be required.

It should be noted that with multiple installations two machines will normally cost less than two single machines as there is common use of some of the associated ancillaries, support structures etc.

A decision needs to be made as to which combination of machines is the best commercial choice.

Conventional availability calculations can be made to show the overall availability of the different configurations, taking into account failure rates, restoration times and scheduled maintenance requirements etc.

The resulting figure for system availability can be related to machine size to give a figure for flow availability . This is shown in Fig. 5 togther with the equivalent number of days per year at 100% production.

An examination of Fig. 5 shows, perhaps not surprisingly, that lowest availability is given by the installations with the lowest installed cost and maximum availability from that with the highest installed cost.

However the "total cost" of any particular system is more than just the installed cost. An unreliable system means that a certain amount of production will not be achieved because of the system downtime. The value of this lost production should be considered as a cost and shown against each system together with the installed cost. Though not included on Fig.5 some evaluations will include operating and maintenance costs in a "total cost". Sensibly it is this "total cost" which determines choice of system and to obtain this it is essential to know the availability of the system.

In order to select the optimum system configuration, a measure needs to be made of the value of any lost production. It is pertinent to observe that "installed costs" and "cost of lost production" are complex figures which need to be supplied by a project's economist. "Installed costs" include direct purchase and installation costs and also the cost of supporting steel work; control system; fuel supplies; fire protection systems; etc.

The cost of a day's lost production is also complex to calculate. Oil Production is not actually lost but it is rather delayed to the end of the field life. This delay can be costed, it is greatly influenced by the tax regimes that are peculiar to offshore oil production and to the production plateau of the field.

The various costs can and must be derived for the commercial viability of the project to be assessed and the optimum design identified.

For the example considered in Fig. 5, a day's "lost" production that will occur each year of plateau life is assumed to be worth £4mm, ie. the difference between producing at 100% throughput for 364 instead of 365 days each year is £4mm over the life of the field.

Multiplying this figure by the number of days per year equivalent that the system is not available enables the loss associated with each equipment configuration to be calculated. This is shown as "production loss" in Fig. 5.

It is now possible to show the "total cost" of each system configuration by summing the installed cost and the production loss. The example shows that the lowest "total cost" is given by the 3 x 50% configuration and this is the "optimum" choice.

In real life it is probable that serious attention would also be given to the 3 x 65%, 3 x 75% and 2 x 100% configurations.

An actual study would also have to recognise that a nominal 50% capacity machine may be able to support say 60% oil production for short periods, as during times of a machine failure non-critical operations are stopped until power generation is again at full capacity.

Similar evaluations can be made for all the major systems on the platform.

The same type of study applied to a platform producing gas for direct sale to a gas supplier would require a different evaluation. The penalties for lost production are usually more immediate as a penalty is applied for failure to supply the agreed volume of gas on any particular day. The number of days when 100% production cannot be achieved and duration of shutdowns thus becomes more important.

It must be realised however that availability, though a major influence in manufacturer's deciding choice of equipment will not be the sole factor in determing which machine is finally purchased, many other variables as delivery time, price, weight, maintenance requirements, compatability with existing plant etc. will also be taken into account.

4. Data

The validity of the data used in the calculations is of paramount

importance.
The sources of process equipment data open to the availability engineer are at the moment limited and poor.
Figure 6 shows schematically the data that can be used in a study.
The data most relevant for any study is usally that taken from the clients own sites. But site data can be surprisingly time consuming and difficult to collect. It is rarely if ever immediately obtainable in the form that the availability engineer needs it. The data has invariably been collected for other purposes concerned with operating, maintenance or production needs and the type of information recorded and its accuracy will be at a level that was sufficient for the original purpose and not necessarily good enough for the new.
The main source of equipment failure and repair data will normally be the site maintenance records. On a modern plant these are usually computerised and it is a comparatively simple task to extract the required information from them. The accuracy of the information must be checked however as for many reasons the interpretation of failure and the categorisation of type of failure will differ from plant to plant and even depending on which person actually recorded the event in question.
This does not imply carlessness or inefficiency on the part of the operator but reflects the fact that a record that is suitable for one purpose is not necessarily immediately suitable for another.
On older plant, records will not be computerised, and the engineer will have to examine handwritten operating and production logs and inspection records as well as maintenance reports and diaries in order to extract historical failure rates and repair times. Even when carrying out a study on the same plant this data may not be valid for a new operating regime. Repair times may shorten dramatically if a change is made from single shift to double or triple shift working. Failure rates could increase if an old machine is called upon to work at a sustained higher output. These changes will have to be taken into account before the historic site data can be used to assess the availability of the plant in the future. Site records frequently will not refer to the equipment that is to be used in a new design and will usually represent too small a sample to give meaningful information.
Site data does however automatically take into account the company's policies regading operations, maintenance, sparing, stores etc. Most importantly it can be thoroughly analysed and verified to show its relevence to the work in hand.
For all the above reasons, site records whilst being very desirable are rarely a complete source of all the data needed in a study, and data from other sources will be required to supplement it.
A number of commercial data bases are available, amongst the most well known in the UK being the Systems Reliability Data Bank (SYREL) and the more recent HARIS data service. (Ref 1). These are useful but being generic data banks, difficulties can be experienced in verifying the

data source and determining the apparently minor points that can have a large influence on the suitability of the data for the situation under review.

The same problems can be experienced with the generic data published in various text books and in specialist papers and periodicals. Some of the latter however can be very useful if they cover a subject in detail.

Certain publications as the IEEE data handbook (Ref 6) and the Telecom handbook (Ref 7) etc. are published as data sources and indicate the taxonomy of the fields covered and how the data has been collected and collated.

An excellent explanation of how taxonomies need to be developed and data handled is given in the early chapters of the OREDA handbook, (Ref 8). The data in the handbook is widely used and has been compiled from the records of a number of European offshore oil operating companies. The data in the handbook is however in some instances taken from small populations of equipment with short operating lives and has to be handled carefully because of this.

A more recent data collection exercise carried out by the OREDA project has produced an event data base but at the present time this is available to participating companies only.

Most equipment manufactures either cannot or will not release meaningful availability data and are rarely a useful source of data.

If data is not available then figures for failure rates can be derived by using Failure Mode and Effect Analytical techniques (FMEA). (Ref 1). This can be very time consuming and requires site and equipment knowledge. The method can be very useful however if a completely new design of equipment is being used or if data is not obtainable from any other source.

In total the engineer carrying out availability or reliability studies for offshore installations will experience difficulty and frustration in trying to obtain suitable data, but the situation is slowly improving. As the difficulties in obtaining data become apparent and a study may be criticised because of the standard of its data, it is perhaps salutary to reflect how many engineering decisions are made on data that must on examination be classed as very imprecise or unverified.

REFERENCES

1) The Reliability of Mechanical Systems Institute of Mechanical Engineers Guide for the Process Industries (London).

2) Reliability Technology - A.E. Green and A.J. Bourne (John Wiley and Sons).

3) Reliability and Maintainability in Perspective - D.J. Smith (The MacMillan Press Ltd).

4) Practical Reliability Engineering - P.D. T.O'Connor (John Wiley and Sons).

5) Slochastic Decisions - Engineering Design - P.E. Simmons
Seminar Paper - Cost Effective Reliability in Mechanical Engineering - Institute of Mechanical Engineers (November 1988) (London).

6) IEEE Guide to the Collection and Presentation of Electrical, Electronic, Sensing Component, and Mechanical Equipment Reliability data for Nuclear Power Generating Stations (Willey Interscience).

7) Handbook of Reliability Data for Components used in Telecommunication Systems - British Telecom (London Information (Rowse Muir) Ltd).

8) OREDA - Offshore Reliability Data Handbook (Distributed by Veritec. P.O. Box 370, N-1322 Hovik, Norway).

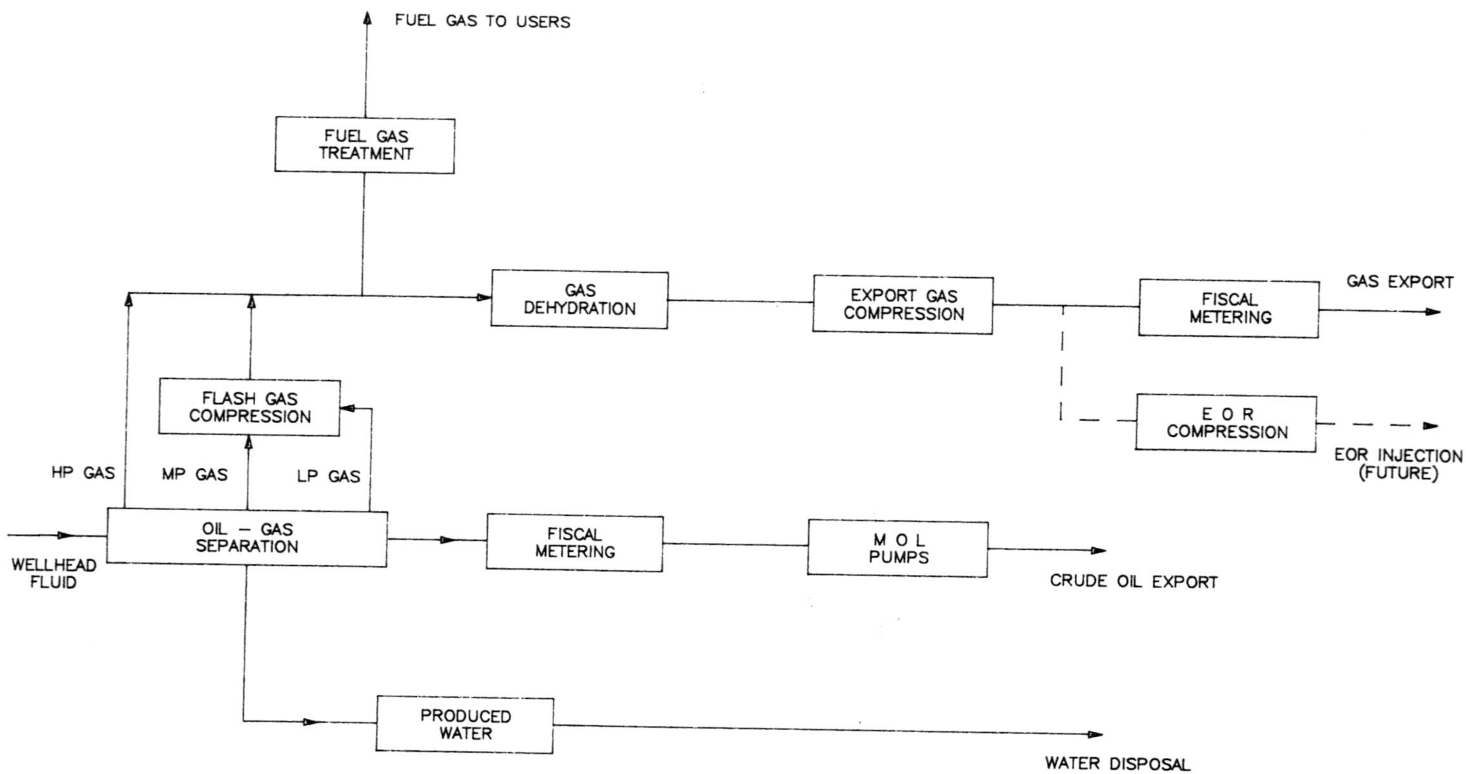

Process Flow Diagram (Schematic)

Fig. 1

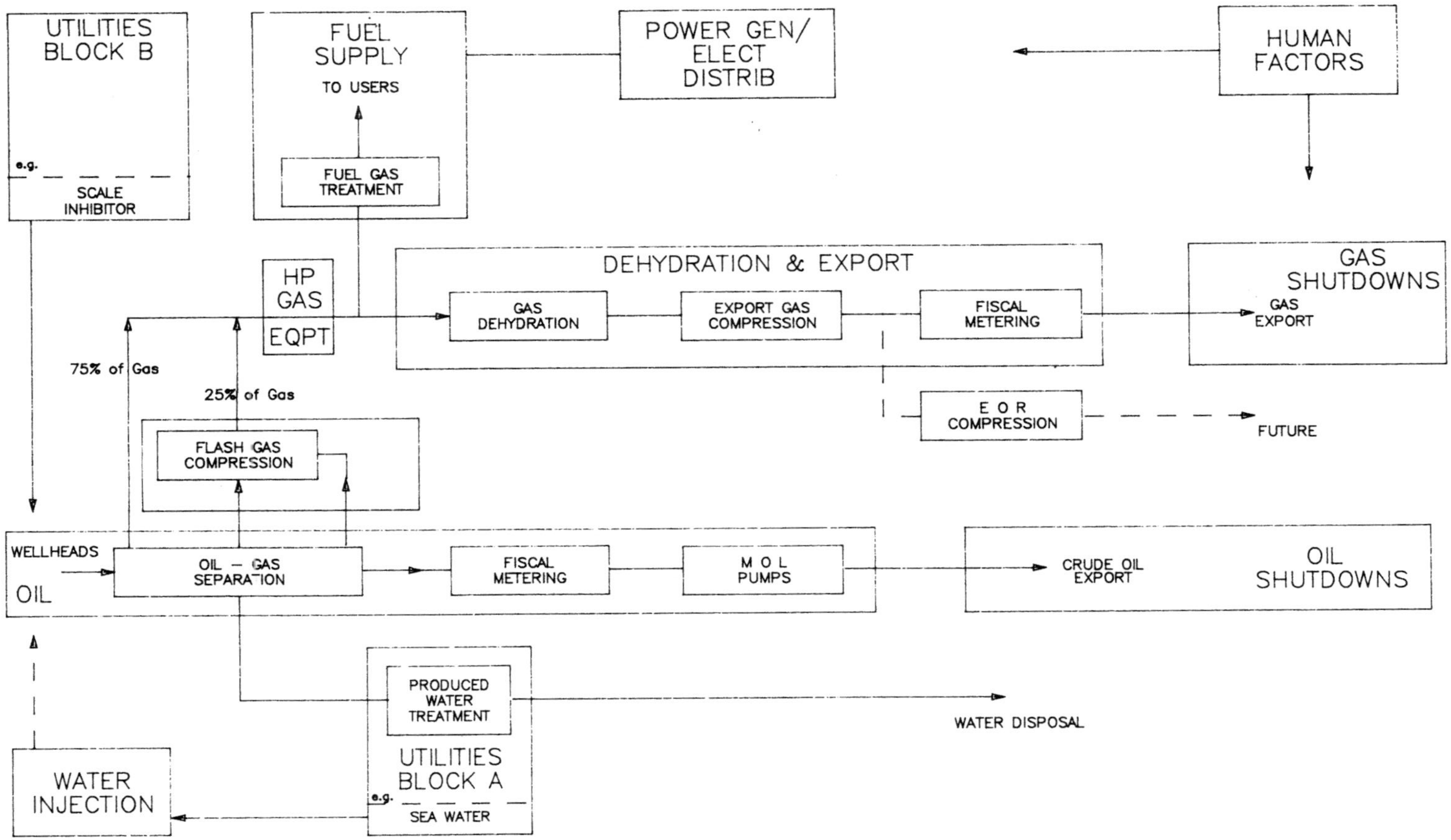

Availability & Process Block Diagram

Fig. 2

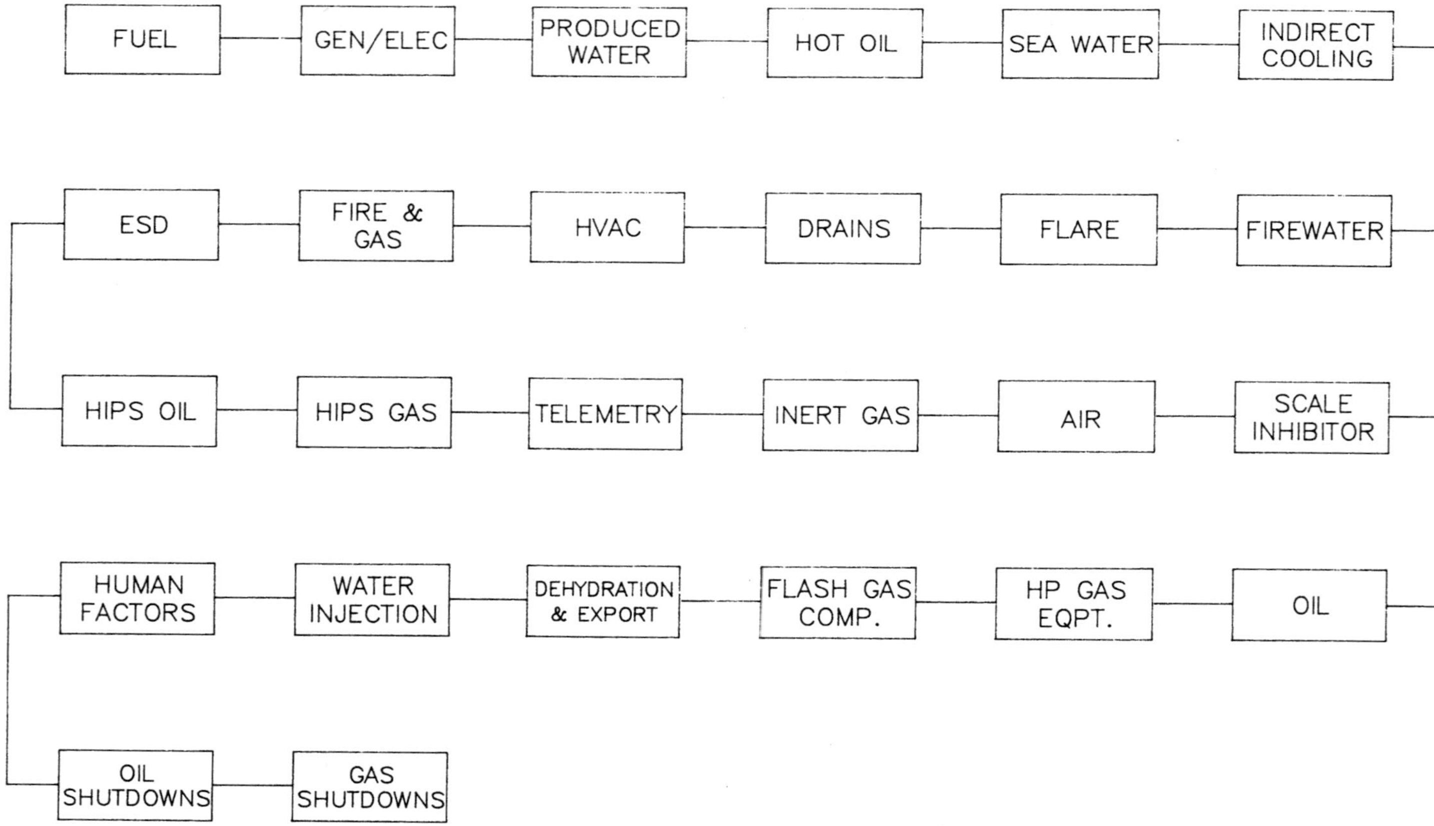

Availability Block Diagram (Utilities)

Fig. 3

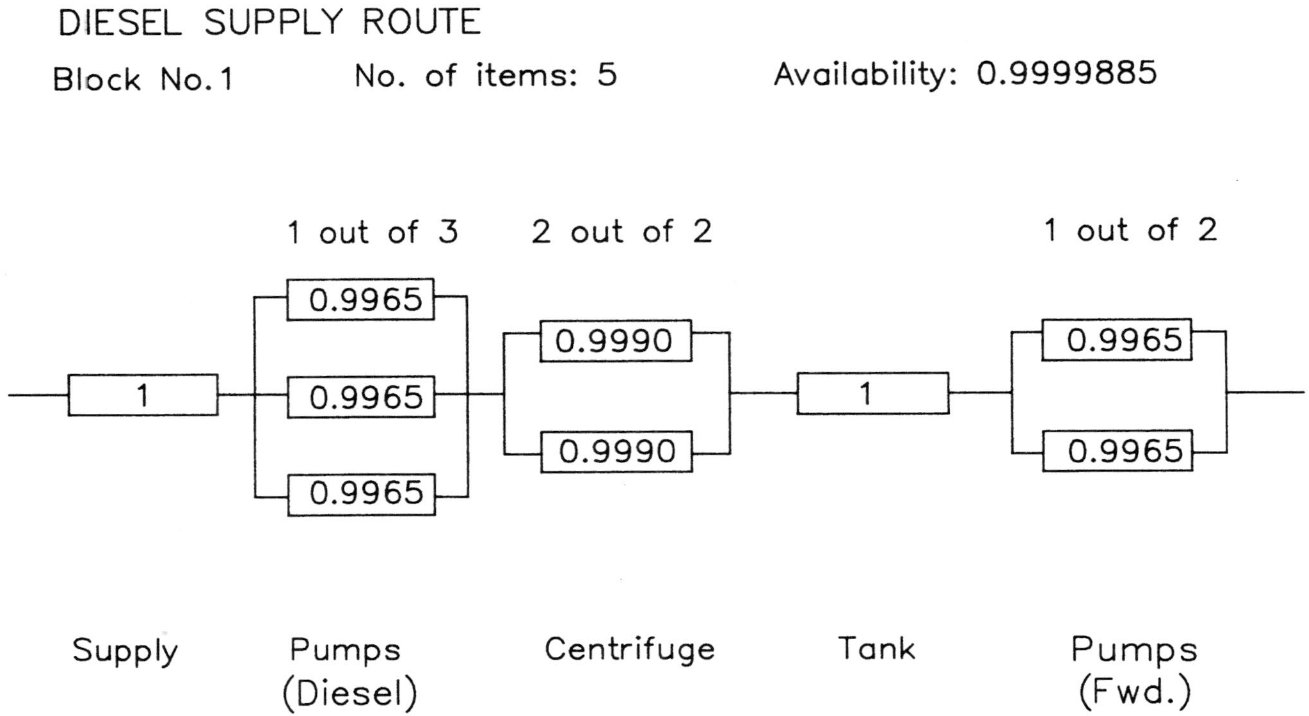

Availability Block Diagram (Detail)

Fig.4

System Configuration	Installed Cost (£millions)	System Availability	Equivalent Flow Availability	Production Loss (£millions)	"Total Cost" (£millions)
1x100%	13.0	.9718	354.7	41.2	54.2
2x50%	15.0	.9718	354.7	41.2	56,2
2x65%	16.0	.9805	357.8	28.8	44.8
2x75%	17.0	.9859	359.8	20.8	37.8
2x100%	25.0	.9992	364.5	1.2	26.2
3x50%	21.0	.9989	364.5	2.0	23.0
3x65%	22.5	.9992	364.7	1.2	23.7
3x75%	24.0	.9993	364.8	0.8	24.8
3x100%	36.0	1.000	365	0.0	36.0

Cost Comparisons

Fig.5

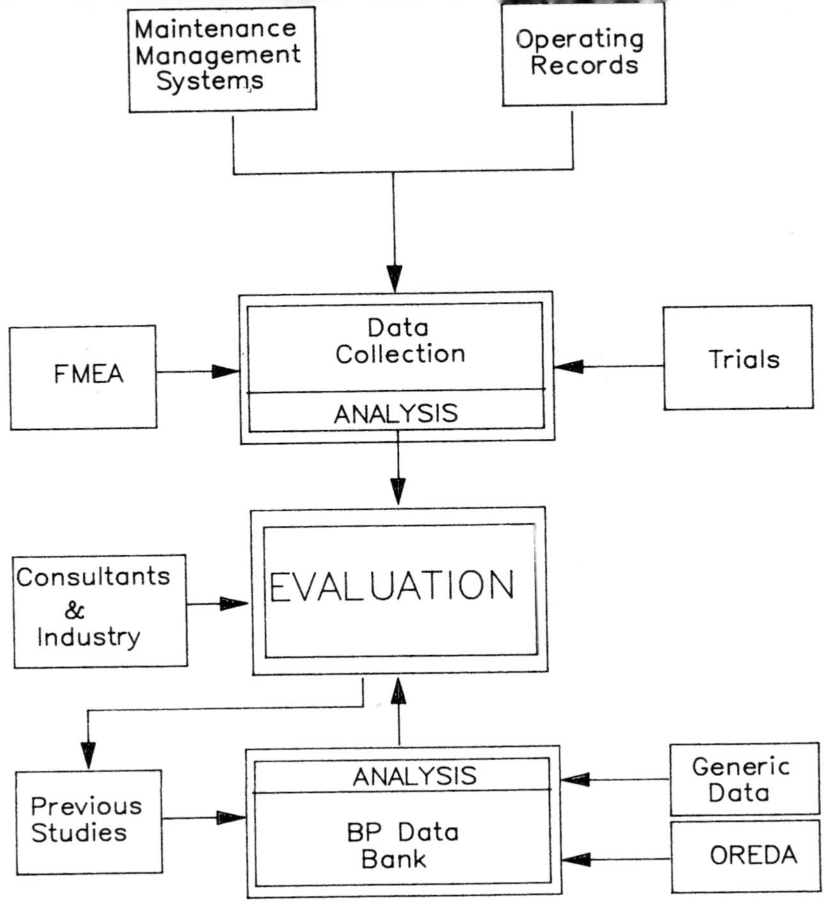

DATA SOURCES

Fig.6

AN ANALYSIS OF ACCIDENTS WITH CASUALTIES IN THE CHEMICAL INDUSTRY BASED ON THE HISTORICAL FACTS

Ing. L.J.B. KOEHORST
TNO Division of Technology for Society Department of Industrial Safety

1. Preface

On behalf of the Directorate-General of Labour, TNO carried out an analysis of accidents with casualties which occurred in the chemical industry.
From the databank FACTS, 700 accidents have been selected. The selection was focused on accidents during industrial activities with hazardous materials. The aim of the analysis was to find out how casualties occur and to give concrete suggestions and recommendations to improve safety during daily practice. The first results are presented in this paper. By the end of 1988 the full report of the analysis will be submitted to the Directorate-General of Labour, and most probable it will become available as a publication of the Directorate.

2. Scope of the analysis

The 700 accidents were selected from the databank FACTS.
The following selection criteria have been used:
- accidents with casualties;
- accidents with hazardous materials, which happened in the chemical industry;
- accidents that happened during the following activities;
- processing;
- storage;
- industrial use and/or application;
- transhipment to or from storage tanks.

J. Flamm and T. Luisi (eds.), Reliability Data Collection and Analysis, 161–180.

Only accidents that were in line with all three criteria were selected.
Figure 1 represents the activities during which the accidents occurred.

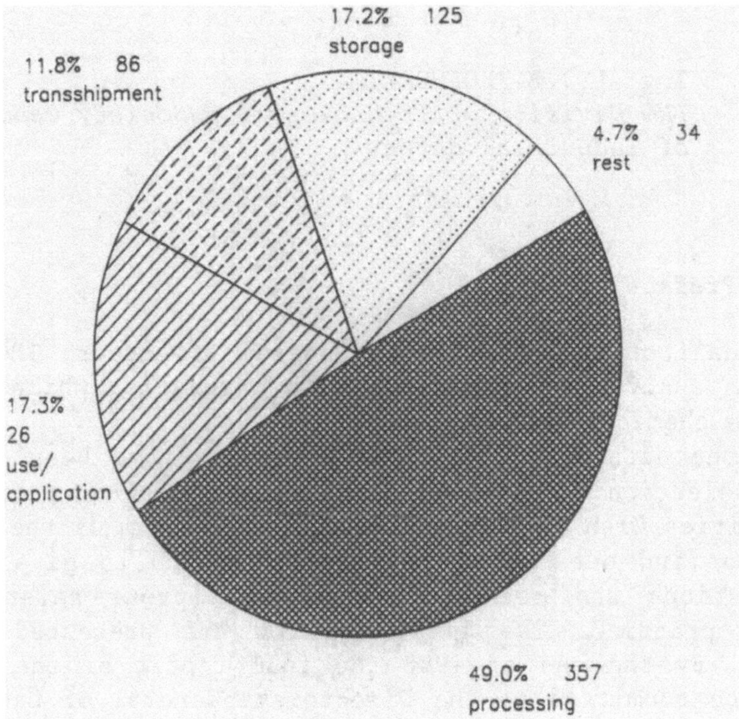

Fig.1. Accidents during several activities

3. The databank FACTS

FACTS is a database with technical information about accidents with hazardous materials that happened during all types of industrial activities (processing, storage, transhipment, transport, etc.). At this moment FACTS contains information of about 15,000 accidents that happened all over the world. Most information concerns accidents during the last 30 years.
The information stored in FACTS is derived from several sources such as:
- literature;
- periodicals;

- technical reports;
- environmental and labour inspectorates;
- industrial companies;
- fire brigades;
- police.

There will always be a discrepancy between the number or accidents that actually have happened and those that are recorded. Accidents having minor consequences may not have been recorded at all, while accidents with more consequences may be recorded incidentally. Only accidents where severe damage or danger is involved will be publicized, analysed and documented. The discrepancy between events that actually have occurred and those that have been recorded, is shown in figure 2.

The quality of the available information of recorded accidents is also related to their seriousness. The most serious accidents are also those of which good and detailed information is available.

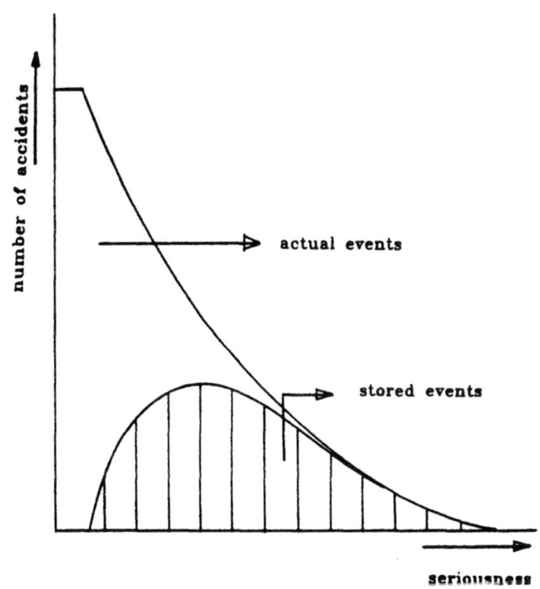

Fig. 2 Comparison of actually occured incidents and recorded incidents

4. Analysis

The selected accidents are divided into three categories:
- normal operation;
- maintenance;
- start-up/shut-down.

Each category represents one specific phase of the lifecycle of an installation or plant. During each phase certain activities will be carried out or chemicals and equipment will be used, which might be of influence on the cause and consequences of accidents. Each phase is analysed with respect to the following items:
- the course of the accident;
- the accident cause;
- type of involved equipment;
- human handling.

Figure 3 illustrates during which phase the accidents took place.

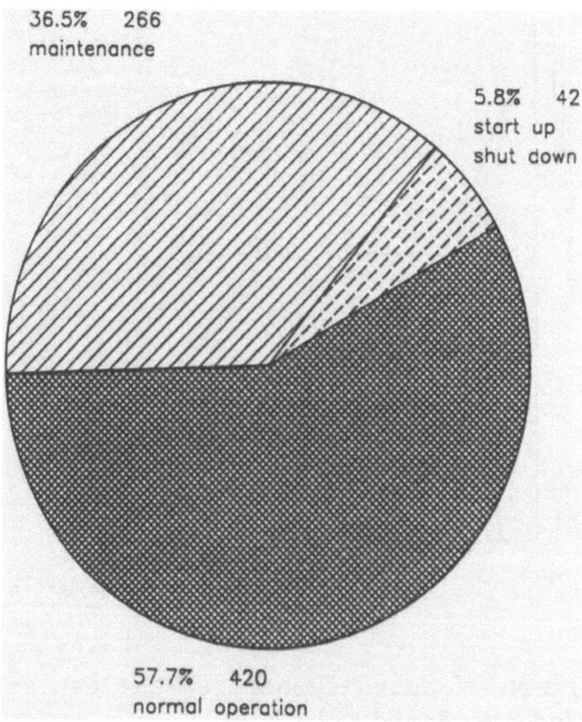

Fig. 3. Classification of the selected accidents into system phases

It is obvious that most accidents took place during normal operation. Remarkable, however, is the great amount of accidents during maintenance and the very few accidents during start up and shut down procedures. With respect to the accidents during maintenance it should be noticed that maintenance time is only a small part compared to normal operation time. This indicates that the probability that accidents occur is relatively high during maintenance.

4.1 The course of accidents

During all three system phases a lot of dangerous situations were created by human actions. In order to describe the consequences of these actions and to indicate their relationship with other events that happen during an accident, the course of accidents has been analysed. The results are presented in a scheme. It will give an insight into the mechanism which leads to the release of chemicals. In the scheme the number placed before each event indicates how many times the event occurred. To keep the schemes readable some occurrences of the same type are replaced by a more general expression. For example,

wrong physical conditions: under-/overpressure;
under-/overheating;
overfilling.

technical failure: - malfunctioning of equipment;
- broken equipment;
- wrong specifications of equipment.

In the figures 4, 5 and 6 phase an event-scheme is presented for each system phase. As might be expected a lot of accidents occur during human handling. This applies for all three system phases. In more detail, it can be concluded that:

* during normal operation a lot of accidents (23%) occur during unloading or pumping over of chemicals;
* 70% of the maintenance accidents occur during dismantlement of equipment (28%), welding (20%) and repair of equipment (20%);
* during start-up/shut-down, accidents often occur during manipulation of equipment (20%) and during dismantlement

of equipment (19%).

All these different types of human action during which accidents occur can be described as simple human actions while people stay in close or direct contact with hazardous chemicals.

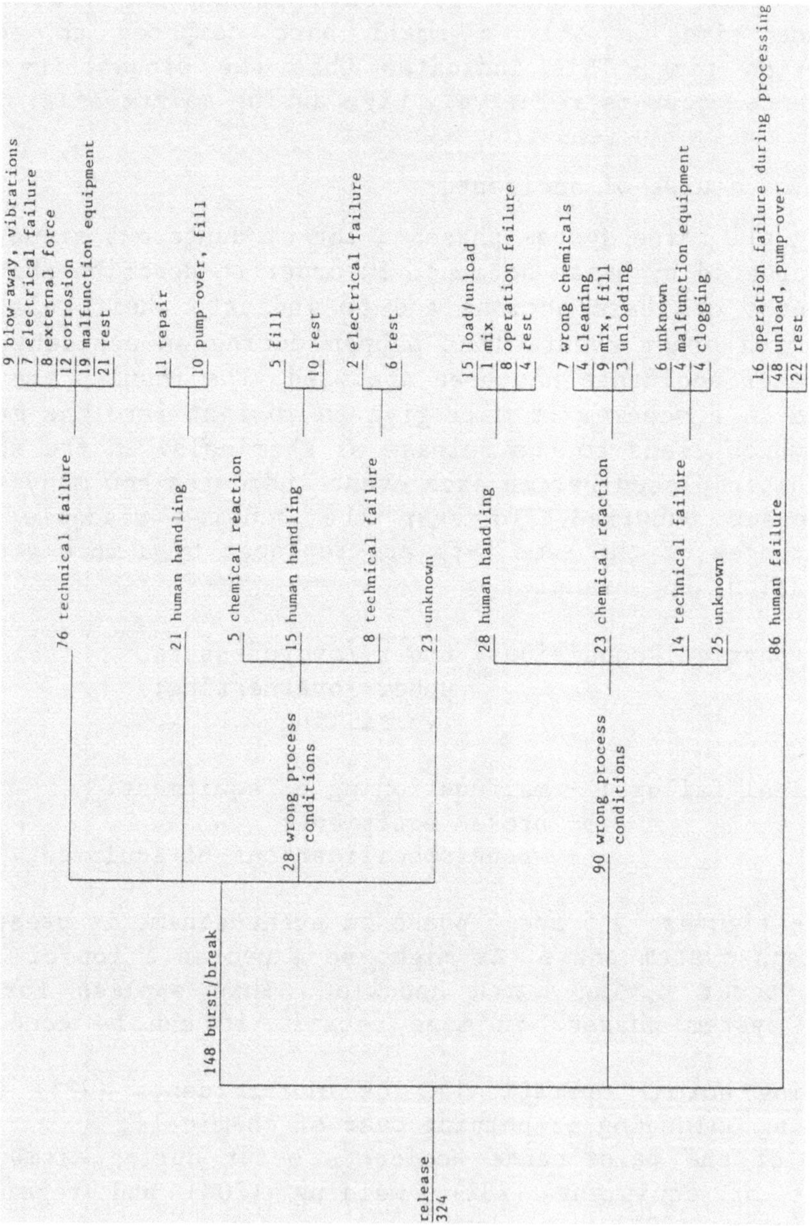

Fig. 4. Event scheme for accidents during normal operation

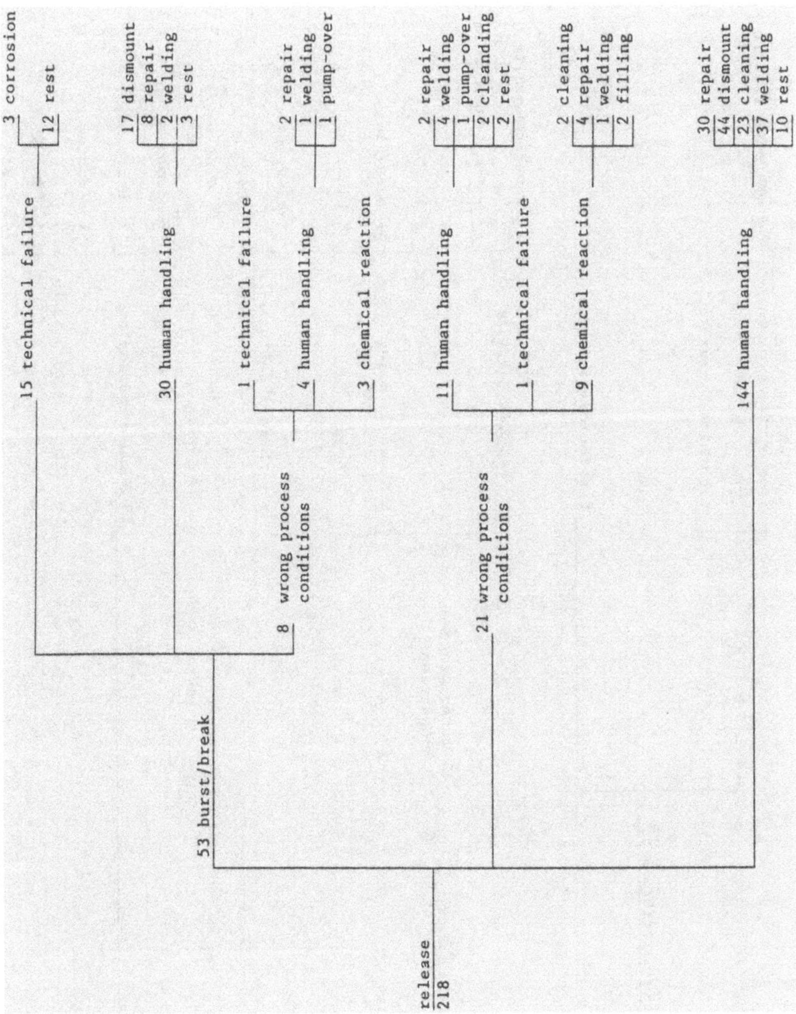

Fig. 5. Event scheme for accidents during maintenance

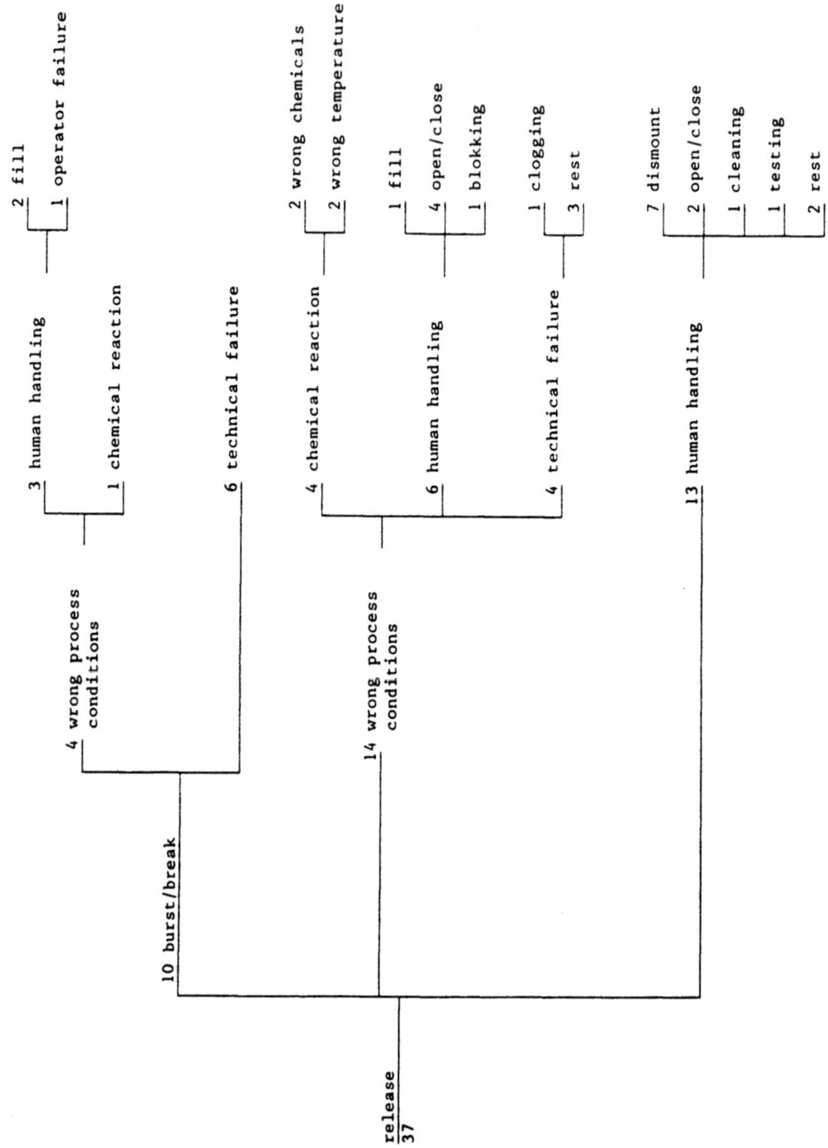

Fig. 6. Event scheme for accidents during start up/shut down

Almost 80% of all selected accidents result into a release of chemicals. It is not possible to indicate which specific chemicals or groups of chemicals are mostly involved. Because most chemicals used in the chemical industry are highly inflammable, a release is often followed by a fire or explosion. Figure 7 presents the type of ignition source involved. Remarkable here is, that in 32% of the accidents during maintenance, welding was identified as the ignition source.

ignition sources	normal operation %	maintenance	start-up/ shut-down
hot spots	8	11	21
electrical sparks	24	3	16
static electricity	8	10	-
open fire	28	20	32
mechanical sparks	21	21	10
self-ignition	8	3	16
welding	3	32	5
total	100	100	100

Fig.7 Classification of ignition sources

For all three system phases the physical consequences of an accident are illustrated in figure 8. In about 30% of the accidents the consequences are limited to a release. In 60-70% a release is followed by a fire or explosion. Especially during normal operation the probability that a release will result into a fire is most likely.

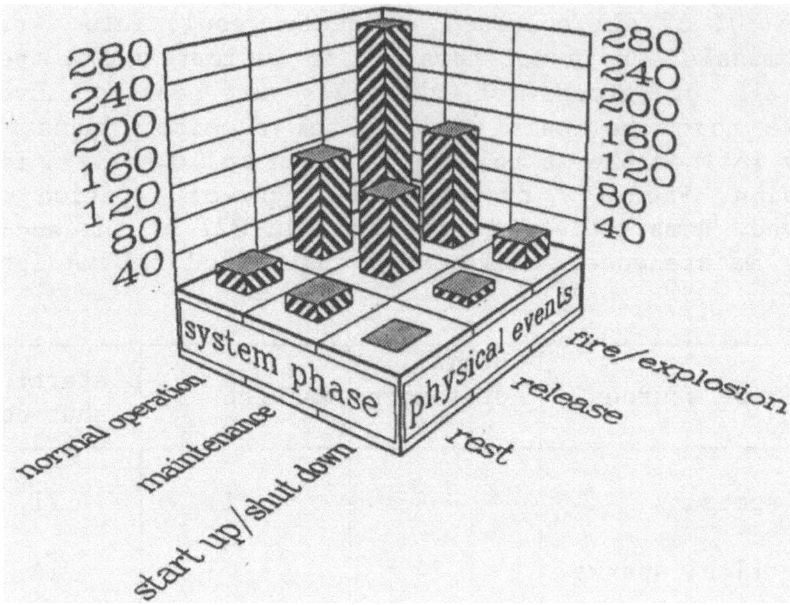

Fig.8 Physical consequences of accidents related to the system phases

4.2 Relation between involved equipment and human handling

With respect to the accidents during normal operation and during maintenance it has been investigated which type of equipment or installation is frequently involved in accidents. From tabel 1 in Annex 1 it can be concluded that during the most accidents the following types of equipment are involved:

- hoses involved in 95 accidents;
- lines involved in 85 accidents;
- valves involved in 152 accidents;
- drums involved in 82 accidents;
- cylinders involved in 58 accidents.

The above mentioned type of equipment are parts of the following installations:

- storage tank/vessel 256 times involved in accidents;
- reactors 72 times involved in accidents;
- transport equipment 134 times involved in accidents.

• In the tables 2, 3 and 4 of Annex 1 the above mentioned

types of equipment and installation are viewed in relation to several human actions. From the event-schemes presented in paragraph 4.1 several human action, which are often involved in accidents, have been selected for this comparison.

- From the tables 2,3 and 4 it can be noticed that a lot of accidents happen in the following situations.

1 Normal operation:
  - handling of valves;
  - use of hoses during loading or filling;
  - filling of drums; regular accidents occur, caused by unclean drums or by using the wrong chemicals. In both situations unforeseen chemical reactions are generated;
  - filling of reactors and tanks; overfilling takes place due to inattention of the responsible persons or to failure of a level indicator.

2 Maintenance:
  - dismantling or disconnecting of lines, valves and hoses, connected to tanks or reactors;
  - cleaning of storage tanks;
  - use and repair of valves; wrong valves are open or closed causing the release of chemicals. Releases also occur from pressurized systems where valves need repair. In several cases the system was not depressurized before repair was started.

3 Start-up/shut-down.
  Although the available number of accidents that took place during start-up or shut-down is rather small, it can be noticed that most incidents were caused by simple human actions, such as handling of valves and lines.

5. Consequences to involved people

Taking in mind the fact that only accidents were selected where people got injured, figure 8 presents the number of accidents related to a cumulative number of injuries and fatalaties. A common rule is that the ratio between fatalities/injuries is about 1:30.
Figure 8 indicates a ratio of about 1:3 à 5. The reason for

this discrepancy is the fact that for this analysis accidents are selected with more serious consequences. This was necessary because these accidents are mostly well documented, which is necessary for such an analysis.

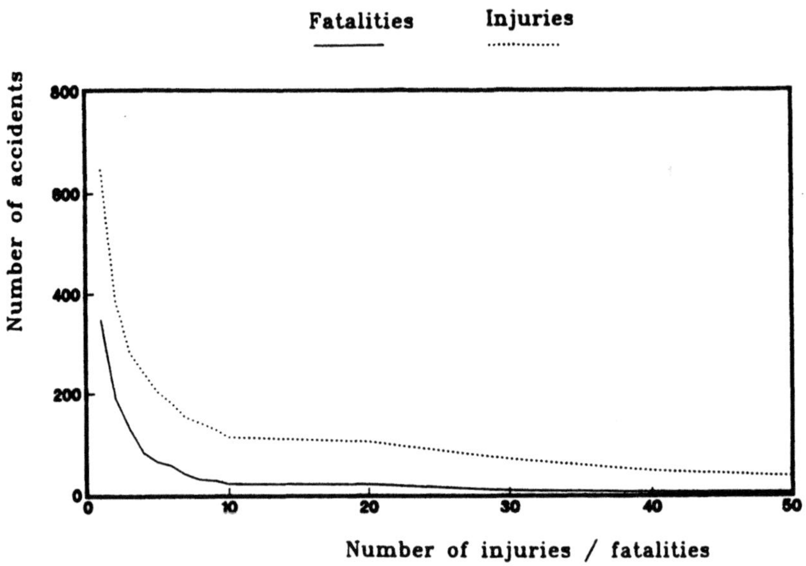

Fig.9 Cumulative number of fatalities/injuries.

Referring to the three system phases, figure 9 presents the cause of injury, related to the number of accidents. The figure indicates that during all system phases, fire is responsible for most of the injuries. During normal operation, also many injuries occurred due to contact with toxic compounds.

Further investigation showed that the numbers of people who were injured by fire or by toxic chemicals were about the same. Toxic chemicals can be present in process installations but can also be formed during a fire of non-toxic substances (plastics, synthetic material etc.). In about 10% of the accidents with a fire, toxic chemicals were formed by which people became injured.

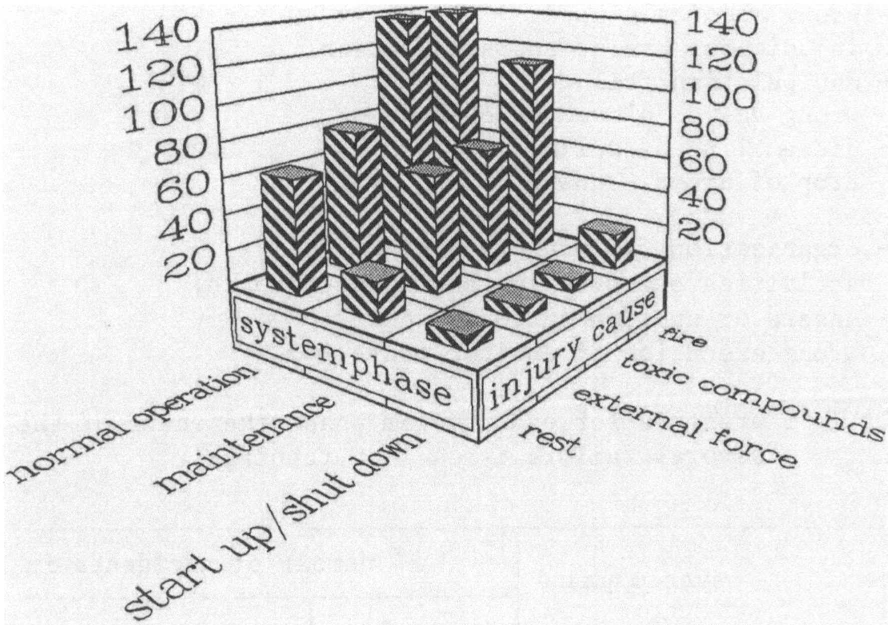

Fig.10 Injury cause related to the system phases

6. Human role during accidents

From chapter 4 it become clear that regardless the system phase, human behaviour is of great influence on the cause and course of accidents. For this reason more specific attention was payed to this subject. In order to qualify the human role during accidents, the following classification has been made of human failure:

- operator failure;
- manipulation failure;
- installation failure;
- inspection failure;
- maintenance failure;
- assembling failure;
- organization failure.

A few examples of some of these failure types illustrate their meaning.
- operator failure:
- wrong process adjustment;

- wrong chemicals used;
- injudicious use of an installation.
- manipulation failure;
- wrong valves closed or opened;
- dismantling of a line;
- drop of drums, cans etc.

- organization failures;
- activities executed in the wrong sequence;
- unsafe or unclean surroundings;
- wrong execution of another man's task

Table 5 presents for each system phase the share of the several failure types in percentages.

system phase / failure type	number of accidents in %		
	normal operation	maintenance	start-up/ shut-down
operator failure	34	28	29
manipulation failure	31	27	29
installation failure	12	1	10
inspection failure	3	6	4
maintenance failure	5	8	14
organization failure	15	30	14

Operator, manipulation and organization failures frequently cause accidents. Examples of such failures that frequently occur are:

* Operator failure:
  - wrong adjustment of an installation that deviates from the optimal setting;
  - the use of wrong combinations, quantities or type of chemical;

- deviation of instructions or specifications, during loading or unloading. No earth wire was connected or the loading was done with too high a velocity, which causes a static electrical discharge.

* Manipulation failures:
  - replacement of valves, gaskets at pressurized or filled installations;
  - opening or closing the wrong valves; dismantling the wrong line or hose;
  - unpermitted smoking, unallowed entrance of an installation, fail to use personal protection equipment.

* organization failure:
  - accidents caused by a combination of activities which are executed at the same time e.g. dismantling a valve and testing a pump in the same system.
  - working in unsafe areas, e.g. vessels which have not been degased, welding in poor ventilated rooms:
  - fail to instruct contractors, or temporary employees.

7. Conclusions and recommendations

In the previous chapters several aspects of accidents with casualities have been analysed and where possible conclusions have been made.
The final conclusion of this analysis, is that most accidents are caused by failures during simple human actions. On the one side this conclusion may seem obvious because especially simple handlings occur very frequently. On the other side just because the high frequency of these handlings the people in charge are not (anymore) aware of the potential danger of chemicals and equipment they use.
Because it will never be possible to be 100% sure that these types of failure will be avoided, further investigation should be aimed at the improvement of safety in these situations. A possibility is the development of a working environment, where the consequence of failures will be limited to a minimum. In analogy to process installations which are inherently safe, this philosophy should lead to a inherently safe working environment.

8. References

[1] Lees, F.P.:
Loss Prevention in the process industries, 1980.

[2] Bockholts, P. and Koehorst, L.J.B.:
Accident analysis of vapour cloud explosions, 17th June 1987.

[3] Bockholts, P.:
Oorzaakanalyse,
1 november 1985.

[4] Oorzakenboek (protoversie),
DGA/TNO, 1987.

[5] Health and Safety Executive.
Dangerous Maintenance.
ISBN 0 11 883957 8, 1987.

[6] Gevaren van statische elektriciteit in de procesindustrie.
Rapport van de stuurgroep Richtlijnen Veiligheid Procesindustrie,
1975.

Annex 1

Table 1

equipment	number of accidents	
	normal operation	maintenance
bots, nuts	8	15
drains	5	9
filters	6	4
fittings	11	10
fanges	4	21
hoses	40	54
lines	35	50
packings	9	6
pumps	20	23
valves	75	77
measure and controle devices	9	6
heat exchangers	7	17
compressors	4	5
drums	71	11
cylinders	41	17

installations	number of accidents	
	normal operation	maintenance
gasholder	1	7
reactors	50	22
tanks/vessels	150	106
transport equipment*	82	52
destillation units	10	3
flares	4	4
furnaces	10	4

* for example tankwagons, railwagons, vessels etc.

Annex 1

Table 2 Accidents during normal operation; human handling viewed in relation to type of equipment/ installation

Human handling Equipment/ Installation	fill	load/ unload	collision	cleaning	pump over	actions: start, stop, open, close etc.	Total number of accidents
hoses	10	9	4	2	-	4	40
pipelines	2	-	5	3	1	6	35
valves	5	7	1	4	1	18	75
drums	25	8	1	1	3	-	71
transport equipment	11	15	6	1	2	6	82
reactors	12	-	-	1	1	9	50
storage tanks	34	11	2	4	5	23	150

Annex 1

Table 3 Accidents during maintenance; human handling viewed in relation to type of equipment/installation

Human handling / Equipment/Installation	cleaning	actions start, stop, open, close etc.	dismount	fill	pump over	load / unload	repair	mount	mainte-nance	welding	Total number of accidents
hoses	6	10	21	6	2	3	9	5	5	6	54
pipelines	6	8	23	-	-	-	6	4	5	8	50
valves	7	23	33		2	1	18	8	2	8	77
drums	2	2	4	1	1	-	3		-		11
cylinders	-	3	4	1	-	-	-	3	1	2	17
transport equipment	5	5	8	2	2	5	3	-	1	8	52
reactors	2	7	10	2	-	-	5	1	3	2	22
storage tanks	27	7	21	7	4	4	14	2	7	15	106

Annex 1

Table 4 Accidents during start-up/shut-down; human handling viewed in relation to type of equipment/ installation

Human handling Equipment/ Installation	open/ close	dismount	blocking	fill	Total number of accidents
valves	6	2	1	-	10
flanges	-	1	-	-	2
packing	-	-	-	-	2
lines	-	3	2	-	10
reactors	-	-	-	2	6
storage tanks	-	-	4	2	8

SYSTEMATIC ANALYSIS AND FEEDBACK OF PLANT DISTURBANCE DATA

KARI LAAKSO, PEKKA PYY, ANTTI LYYTIKÄINEN
Technical Research Centre of Finland (VTT/SÄH)
Laboratory of Electrical Engineering and
Automation Technology
SF-02150 Espoo, Finland

ABSTRACT

This paper contains a description of methods to be used in a systematic analysis of opportunities for reducing the number of unplanned production losses.

The use of the model, developed for analysis of operating experience, ensures that possible measures for significantly reducing the number of plant disturbances will be systematically identified, analyzed and ranked.

Emphasis is given to reduction of the number of inadvertent process disturbances, leading to forced outages. This reduction has also a beneficial effect on plant safety, because the number of possible accident initiators is decreased.

Examples of steps included in the model are given as follows:

- collection of incident reports as well as operational and maintenance reports to be used as source data
- detailed analysis of the event sequences and the contributing causes to the plant disturbances in production plants
- identification and quantification of recurring failure events and their long-term trends at a suitable functional level in individual plants
- analysis and ranking of possible, proven or yet unproven, measures for the plants, based on their expected effectiveness in reducing the number of production disturbances
- presentation and discussion of incident analyses performed and measures recommended with personnel at the plants.

The analysis of process parameters, using a computerized disturbance recording function with a high sampling rate, significantly improves the possibilities to identify the contributing causes to sudded plant incidents.

J. Flamm and T. Luisi (eds.), Reliability Data Collection and Analysis, 181–192.

The model can also be used for comparison of rather similar plants' failure frequencies and trends at the functional group level, in order to identify significant improvements achieved, or opportunities for improvements, to be transferred between the plants. The studies performed at Swedish nuclear power plants resulted in several recommendations, which should significantly reduce the number of unplanned reactor shutdowns and turbine trips and thus reduce the production losses. The recommendations are based on cost/benefit considerations and they resulted in several modifications of equipment as well as improvements of operating and maintenance procedures.

A part of above steps have now also been further developed, and introduced for analysis and experience feedback of different categories of plant incidents, including technical and human elements, at Imatran Voima Power Company's conventional power plants in Finland.

Application of rather similar analyses is also recommended for the specific processes in other complex industries to be used effectively in the fields of design, maintenance, operational and training improvements.

BACKGROUND TO THE INCIDENT ANALYSIS METHODS DEVELOPMENT

The background to the basic study [1] was that the reduction of inadvertent process transients in a nuclear power plant has a beneficial effect on nuclear safety and on unanticipated production losses. Sufficient reason existed therefore to aim at decreasing the plant disturbance frequency even further. Such a reduction of transient frequency would also lead to a reduction of thermally, dynamically and electrically induced stresses that may contribute to leakage or damage in equipment.

The basic model [1, 2] was earlier developed for follow-up and analysis of BWR plant operating experience in Sweden. The nuclear power units studied were of earlier Asea-Atom design and they (Oskarshamn 1 and 2, Ringhals 1, Barsebäck 1 and 2) are owned by Swedish utilities. The study covered an analysis and feedback of a total of 44 years operating experience.

The methods were applied in analysis of plant disturbances leading to reactor shutdowns, turbine trips and generator load rejections in these units.

The basic research project was performed under the auspices of the Swedish Nuclear Power Inspectorate, in close cooperation with Swedish power utilities, within the Engineering Department of Asea-Atom.

A part of the steps in the model have now been further developed, completed and applied for systematic analysis and experience feedback of different categories of plant incidents, including human contribution, at Imatran Voima Power Company's (IVO) conventional power plants in Finland.

An attention in the later study [3] was also given to the sector, where the methods and model now described were applied and developed to concern the production losses, material damages, near-misses and other significant incidents at IVO's conventional power plants.

The joint development work between IVO and the Technical Research Centre of Finland (VTT) concerned the Inkoo coal-fired 1000 MW power plant consisting of four similar 250 MW electricity producing units [3].

Particular emphasis has been laid on finding out the technical and human elements and how to distinguish them in a proper way, so that the results could be used effectively in the fields of operation, maintenance and design, as well as in the safety analyses, quality assurance and training.

METHODS UP TO DATE DEVELOPED AND USED FOR THE ANALYSIS

The first analysis step, in the basic model for feedback of plant disturbance experience in nuclear power plants, was to perform individual incident analyses.

The sequence of power plant disturbances, including the failure functions and human errors contributing to reactor shutdowns, turbine trips and generator load rejections, has been analyzed using the event reports or analyses, operation reports and maintenance reports from the power plants as source data. Additional information, concerning the disturbances occurred and the corrective actions realized or planned, has been collected through discussions at the units. The preliminary event analyses prepared have then been used as source material for discussions with the operational management and personnel.

To be able to systematically identify the contributing causes to the plant disturbances, and to divide the contributing causes into failure types, five different failure types were defined. The failure types are shown in Table I.

The failure types are specified so that each one is matched with one type of corrective action required.

Improved specification, planning, functional testing or follow-up of maintenance, for example, can be needed in improving the human performance contributing to the failure type 2. A poor human performance in turn can be caused by organizational deficiencies.

Implementation of suitable information presentation in control room or improved operating instructions, for example, can be required in eliminating the human errors, included in failure type 5.

That is why the 5th failure type human error has been divided in the later VTT developments into three different classes; Human error, deficient work planning and flaws in information flow.

Table I Failure types.

1. <u>System malfunction</u> - Unsuitable design due to insufficient knowledge of behaviour of process variables or of man-process interface - Insufficient capacity - Poor redundancy
2. <u>Component failure</u> - Component unsuited to the environment - Unreliable component which can be a result of poor preventive maintenance or ageing
3. <u>Inadvertent protection function</u> - Protection function was tripped even though the event would not have caused any damage (if the trip had not occurred)
4. <u>Testing</u> - Intentional trip due to planned test - Unplanned trip initiated during testing
5. <u>Human error</u> - Incorrect, incomplete or unclear operating instructions (procedures) - Deviations from operating instructions

In order to be able to systematically divide the contributing causes, including human errors, to the plant parts and equipment involved, a suitable functional group division was developed for the Swedish BWR nuclear power units.

A rough hierarchical functional division of a complete nuclear power plant is shown in Fig. 1 as follows.

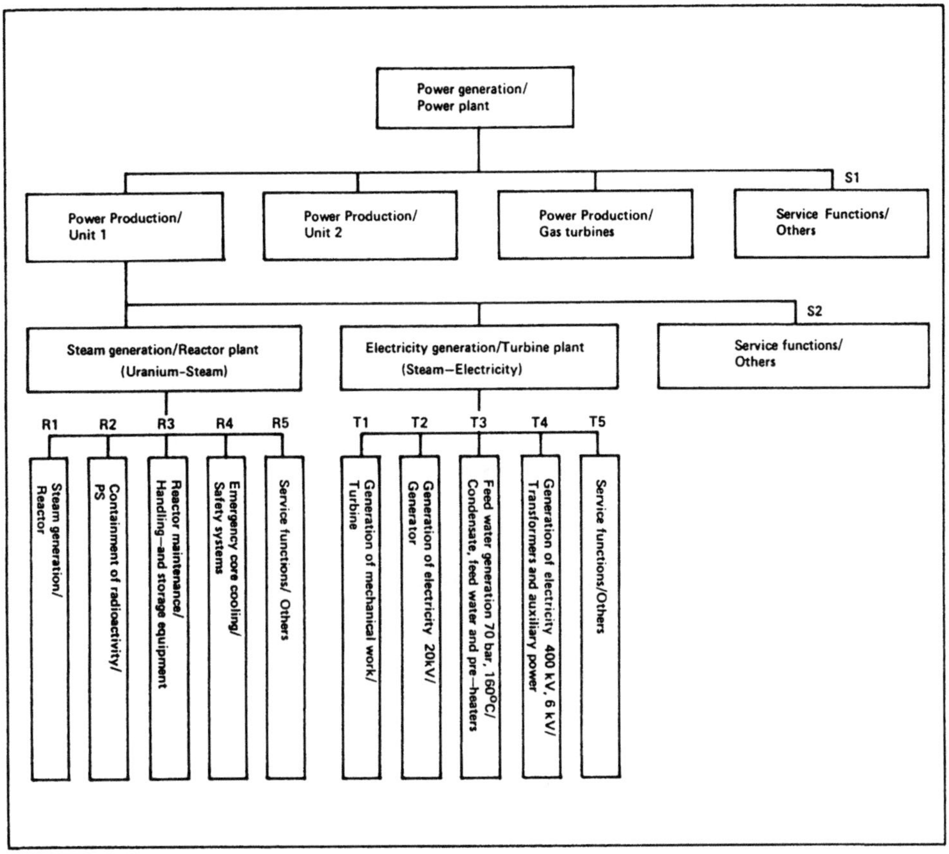

Fig. 1 Breakdown of the power plant into functional groups.

These functional groups (FG) have been made similar for the different units. Therefore these FGs can also easily be used for transfer of operating experience between the different units at the functional group level. Quantification of failure functions and human errors occurred during several incidents, and identification of corrective actions implemented and improvements achieved in individual units, can be performed using trend analyses of these functional groups.

This detailed functional group division is also based on function, and not on the hardware, which makes a functional analysis of the plant disturbances easier than by using the traditional plant system and equipment classifications.

An example of an incident analysis is shown in Fig. 1 as follows.

EVENT ANALYSIS - UNIT 2

BLOCK **DATE** 760714 **TIME** 11.10 **SKI NO** 2-4/76

TRIPPING CONDITION
1:
2: TSS TURBINE TRIP
3: SS10 EXTRA HIGH NEUTRONFLUX RELATIVE TO CORE COOLANT FLOW

	OPERATIONAL DATA BEFORE DISTURBANCE	OPERATIONAL DATA AFTER DISTURBANCE
OPERATIONAL STATE	8	
REACTOR POWER (MWT)	1200	
MC-FLOW (KG/S)	3100	
CONTROL ROD CONFIGURATION	9160	
GENERATOR POWER (MWE)	365	000

FAILURE EVENTS	FUNCTIONAL GROUP	FAILURE TYPE
1. IMBALANCE PROBLEMS OCCURED IN THE TURBINE.	T1:04	1
2. AN ATTEMPT WAS MADE TO COUNTER THE VIBRATIONS, BY INCREASING THE BEARING OIL TEMPERATURE, WITHOUT ANY SUCCESS. ON REDUCING THE TEMPERATUR TO NORMAL OPERATING LEVEL, VIBRATIONS INCREASED FURTHER LEADING TO TURBINE TRIP.	T1:04	4
3. THE MISMATCH OF TURBINE INLET VALVES AND BYPASS VALVES CAUSED A PRESSURE TRANSIENT IN REACTOR CORE.	T1:10	1
4. A "VERY SHORT LIVED" INCREASE OF THE NEUTRON FLUX IN THE CORE RESULTED IN TRIPPING OF REACTOR SCRAM (SS10).	R1:02	3

FUNCTIONAL GROUPS: (FUNCTION/EQUIPMENT)

R1 STEAM GENERATION/REACTOR
R2 CONTAINMENT OF RADIO-ACTIVITY/PS
R3 REACTOR MAINTENANCE/HANDLING AND STORAGE EQUIPMENT
R4 EMERGENCY CORE COOLING/SAFETY SYSTEMS
R5 SERVICE FUNCTIONS/OTHERS

T1 GENERATION OF MECHANICAL WORK/TURBINE
T2 GENERATION OF ELECTRICITY 20KV/GENERATOR
T3 FEED WATER GENERATION 70 BAR,180 C/CONDENSATE,FEED WATER AND PRE-HEATERS
T4 GEN. EL 400 KV, 6 KV /TRANSFORMERS AND AUXILIARY POWER
T5 SERVICE FUNCTIONS/OTHERS

S1,S2 SERVICE FUNCTIONS/OTHERS

FAILURE TYPES:

1 SYSTEM MALFUNCTION
2 COMPONENT FAILURE
3 INADVERTENT PROTECTION FUNCTION
4 TESTING
5 HUMAN ERROR

Fig. 2 A follow-up analysis of incident data. An example.

It should be noticed that one plant incident, e.g. reactor scram (forced shutdown), is usually caused by several interacting failure events and/or functional deficiences in different plant parts and functional groups of the unit. E.g. an automatic scram is seldom caused by a sole failure event only.

In the example incident analysis a human activity is identified as the secondary contributing cause involved in the disturbance sequence. The testing/calibration attempt (failure type 4), performed by the plant personnel, initiated unintentionally the turbine trip, which lead to a reactor scram (forced shutdown).

A computerized disturbance recording, of both analogue and digital process signals, has in many cases facilitated excellent information of the process behaviour during these or similar incidents occurred at another (TVO) nuclear power plant in Finland. The analysis of process parameters, recorded at some units by these means with a very high sampling rate, has thus significantly improved the possibilities to identify and understand the contributing causes, including operator actions, to these sudden disturbances.

The follow-up analyses of incident data, similar to Fig. 2, were commented by and discussed with the personnel at the units prior to the final documentation. All 625 incident analyses, worked out for the earlier BWR-units, were systematically documented in the similar standardized forms, as shown in the Fig. 2 above. Thus storing of all incident analyses in a transient analysis data base was facilitated.

THE MODEL FOR SYSTEMATIC ANALYSIS AND FEEDBACK OF PLANT DISTURBANCE EXPERIENCE

The steps included in the complete model for analysis and evaluation of opportunities, to improve both the plant reliability and the nuclear safety, are summarized in Fig. 3 as follows.

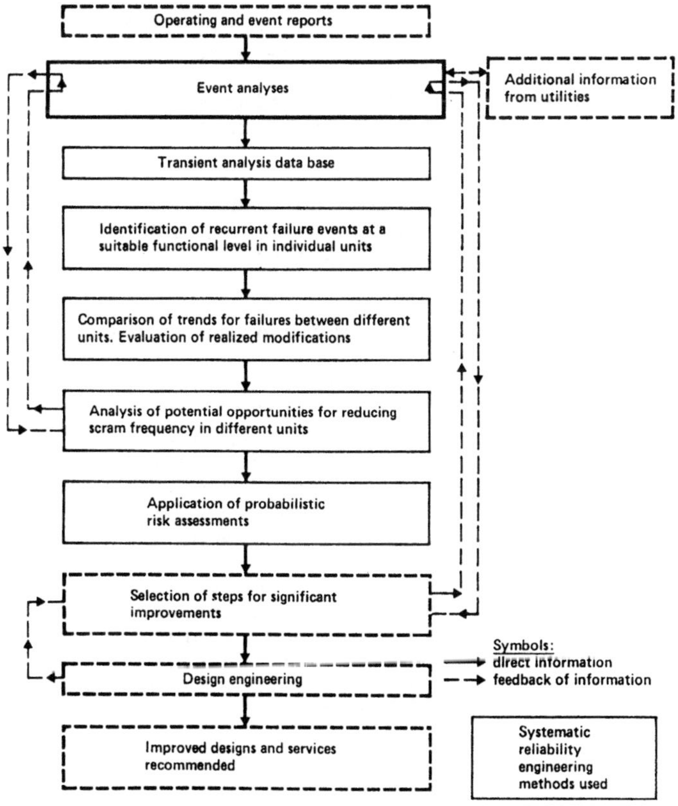

Fig. 3 Steps included in the model for systematic analysis and feedback of plant disturbance experience.

It should be noticed that the effects on nuclear safety, of the proposed corrective measures, can be evaluated by application of probabilistic risk assessments.

It is known that partially similar projects to this, concerning feedback of operating experience or analysis of accident sequence precursors, have been performed in U.S.A. by e.g. the Institute of Nuclear Power Operations, Electric Power Research Institute, Nuclear Regulatory Commission and plant owners and vendors and in other parts of the world. The development work at e.g. the Swedish Nuclear Power Inspectorate [4], the Technical Research Centre of Finland [5, 6] and IVO Power Company [7, 8] has also contributed to tie together the systematic reliability analyses (PSA), the event analyses and human factors analyses in such a way that a systematic feedback is provided to a number of safety activities.

AN EXAMPLE OF SELECTION OF AREAS FOR FURTHER REDUCTIONS OF THE PLANT DISTURBANCE FREQUENCY

Several opportunities for reducing the plant disturbance frequency in individual units have been identified during the earlier analyses performed. As seen e.g. in Fig. 4 below a significant number of failure events, which have contributed to reactor scrams (shutdowns) in Swedish BWRs, originate also from:

- component failures in the turbine plants (which can be a result of e.g. poor planning, performance, testing or follow-up of preventive maintenance).

- operators' human errors (which can depend on e.g. poor instrumentation displays and control equipment or incomplete operating instructions and training).

ESSENTIAL RESULTS AND FURTHER DEVELOPMENTS OR APPLICATIONS

One important result from the work done earlier [1, 2] is that a model and methods, for systematic analysis and feedback of plant disturbance experience, have been developed and used as a powerful tool for both the plant reliability and safety improvement work. This research project [1] was performed under the auspices of the Swedish Nuclear Inspectorate, in close co-operation with the utilities, within the Engineering Department of Asea-Atom.

Several opportunities, for reducing the frequency of reactor scrams (forced shutdowns) and turbine trips, have been identified during this study of plant disturbance experience. Recommendations for improvements in the operating units have been made and they have resulted in modifications of equipment and improvements of operating and maintenance procedures. These improvements were aimed at eliminating either the primary failures or the secondary contributing causes in the disturbance sequences, otherwise leading to tripping of protection functions at plant level and to losses of the electricity production.

A clear experience exists, arising from the earlier study, that

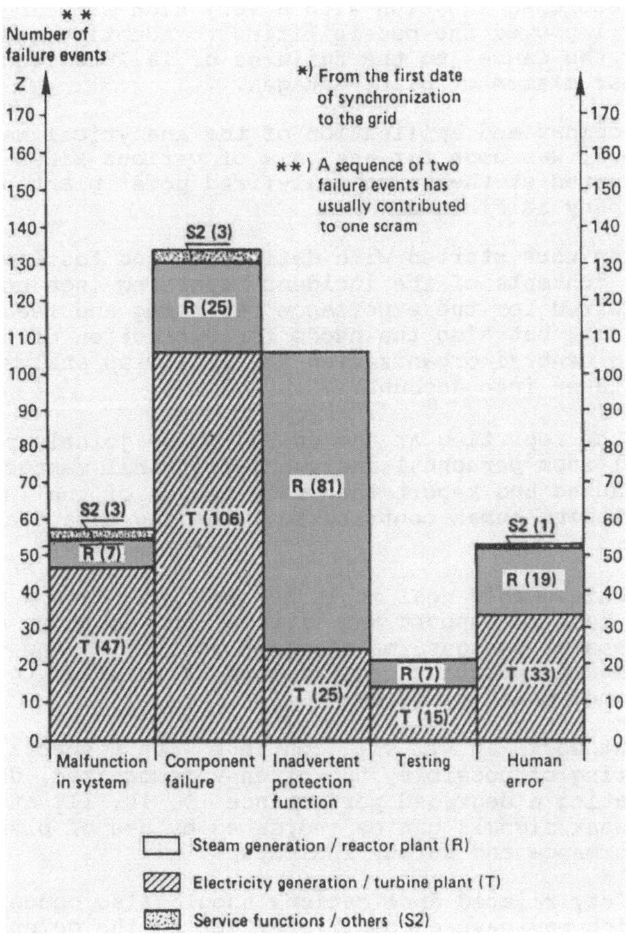

Fig. 4 Reactor scram analysis of a BWR unit. Breakdown of the failure events.

significant reductions of plant disturbances have been achieved, and further reductions can be achieved, with relatively modest efforts and capital costs within systematically selected problem areas. The extra costs for such selected measures (e.g. equipment modifications and improved procedures) in nuclear power units can be compared with expected recurring costs for unplanned power replacement. Viz. in the event of disturbances, a more expensive reserve production from conventional power plants is usually needed.

This analysis also provided an improved understanding of the plant processes, operator actions and maintenance activities, contributing to the experienced disturbances.

The analysis of selected process parameters, using a computerized disturbance recording function with a very high sampling rate, has significantly improved the possibilities to identify and understand rapidly the causes to the failures or failure sequences, leading to near-misses or plant outages.

Further development and application of the analytical methods as presented above, was done for analysis of various kinds of incidents occurred at the Inkoo coal-fired power plant owned by IVO Power Company in Finland.

The application work started with definition and testing of the structure and contents of the incident reporting (see enclosure 1) primarily required for the experience reporting and feedback within the plant, but also the needs for extraction of this reporting to a central organization for follow-up analysis and feedback was taken into account.

One goal for the reporting at the units, to be jointly performed by the control room personnel and the operational management, is to identify causes and report the whole course of the incident sequence, including human contribution, leading to a plant incident.

In order to achieve this goal it is necessary to have a plant incident data base to support and improve the knowledge of e.g. poor and successful designs, maintenance activities and operator actions, which are important contributors in relation to plant disturbances and possible accidents.

The development works at VTT will continue with a special emphasis on the monitoring of possible, but often unrecognized, developments indicating a degraded performance [9, 10, 11] at the units. Such weak signals can be indicated by use of plant-specific performance and safety indicators.

The future safety related applications should also concentrate on incidents, which may have a possible effect on the defence-in-depth principle, where several lines of defence could be made inoperable. Thus a systematic operational safety experience analysis should be applied both backwards in the direction of causes and forwards in the direction of possible consequences [5, 11].

The operating experience of equipment and technical systems, as well as man-process interface, in different complex industrial plants, should preferrably be analyzed in a systematic manner by using these kinds of methods. As a results it should be possible to identify cost-effective measures to improve plant reliability in order to reduce the number of incidents, which cause safety risks, production losses or equipment damage in complex industrial plants.

REFERENCES

1. K. J. Laakso - Systematisk erfarenhetsåterföring av driftstörningar på blocknivå i kärnkraftverk, (A Systematic Feedback of Plant Disturbance Experience in Nuclear Power Plants), Helsinki University of Technology. 1984. Finland. (135 pages in Swedish and 15 pages in English).

2. K. J. Laakso - A Systematic Analysis of Plant Disturbance Experience in Swedish Nuclear Power Units. OECD/NEA Symposium on Reducing Reactor Scram Frequency. April 1986. Tokyo, Japan.

3. O. Viitasaari, U. Vuorio, P. Sammatti, K. Laakso, L. Norros - Ongoing Programs for Analysis of Incident Sequences with Human Contribution at IVO Power Company, IAEA Specialist Meeting on the Human Factor Information Feedback in Nuclear Power. May 1987. Roskilde, Denmark.

4. L. Carlsson, L. Högberg, L. Nordström - Notes on Three Key Tools in Safety Analysis and Development: Incident Analysis, Probabilistic Risk Analysis and Human Factors Analysis, International Conference on Nuclear Power Experience, IAEA, Vienna, 82-09-13--17.

5. B. Wahlström, K. Laakso, E. Lehtinen - Feedback of Experience for Avoiding the Low-Probability Disaster. Presented at the IAEA/OECD/NEA Symposium on the Feedback of Operational Safety Experience from Nuclear Power Plants. Paris, May 1988.

6. P. Pyy, J. Suokas - Identification of Accident Sequences with Human Contribution in the Process Industries. SRE-Symposium -87, Society of Reliability Engineers, Scandinavian Chapter. Helsingør, Denmark, October 1987.

7. L. Illman, J. Isaksson, L. Makkonen. J. K. Vaurio, U. Vuorio - Human Reliability Analysis in Loviisa Probabilistic Safety Analysis. SRE-Symposium. Oct. 1986. Otaniemi, Finland.

8. K. Sjöblom - The Initiating Events in the Loviisa Nuclear Power Plant History. ANS Topical Meeting on Anticipated and Abnormal Transients in Nuclear Power Plants. April 1987. Atlanta, USA.

9. Palmgren - Experience in Plant Performance and Methods for Improving Performance including Refuelling. ENC'86 Conference. June 1986. Geneve.

10. A. Lyytikäinen, O. Viitasaari, R. Päivinen, T. Ristikankare - Improved Management of Electric Switching Stations by Using a Computerized RAM-Data System Module. EuReDatA Symposium, Stockholm, Sweden, Sept. 29th, 1988.

11. J. Holmberg, K. Laakso, E. Lehtinen, U. Pulkkinen - Ongoing Activities of Technical Research Centre of Finland for Nuclear Power Plant's Operational Safety Assessment and Management. Presented at the 2nd TÜV-Workshop on Living-PSA-Application. Hamburg, May 7 - 8, 1990.

VTT/SÄH KL 900724
Enclosure 1

<u>List of Contents for an Incident Report at a Conventional Power Plant</u>
(to be reported directly by the control room personnel in shift and completed by operational management later on)

INCIDENT REPORT

- Report identification number
- Unit's name
- Author(s)
- Date
- Title of incident
- Disturbance initiation time
- Incident termination time
- Plant operational state prior to incident (classification)
- Significant process parameters and other information prior to incident
- Tripping condition(s) (classification)
- Detailed description of course of the event sequence (from first deviation identified up to and including original power level)
- Failure causes (classification)
- Description of corrective actions implemented or proposed
- Functional groups (classification)
- Operational difficulties in management of the event and lessons learned
- Extra costs incl. costs for energy replacement
- Enclosures (Alarm lists, process and parameter recordings)
- Reference to other investigations incl. failure reporting at component level
- Checked by/Date

Obs. The primary emphasis is placed on the describing text, but the text is completed with a classification for explanatory and statistical follow-up analysis purposes.

PROCEDURES FOR USING EXPERT JUDGMENT IN RISK ANALYSIS

R. M. COOKE
Department of Mathematics and Informatics
Delft University of Technology
PO Box 5031
2600 *GA Delft*
The Netherlands

INTRODUCTION AND SCOPE

This chapter describes step by step procedures for the development of expert judgment data for probabilistic risk analysis. Many models for combining expert judgments can be found in the literature (for a survey see Cooke, 1991). However, consistent with recommendations to the Dutch Ministery of Housing, Physical Planning and Environment, and to the European Space Agency, only the so called Classical Model is recommended for implementation at this time. The development of expert judgment data is broken into nine phases, which are addressed in the corresponding sections of this chapter:

1 **The problem identification phase**
2. **The expert identification phase**
3. **The expert choice phase**
4. **The question formulation phase**
5. **The calibration variable selection phase**
6. **The elicitation phase**
7. **The combination phase**
8. **The discrepancy analysis / feedback phase**
9. **The documentation phase**

Broadly speaking, a probabilistic risk assessment can be broken into two parts. One part, which may be denoted **accident prediction**, concerns the assessment of the occurrence rates of undesired events (sometimes called Top Events). The dominant methodology in this phase is fault tree analysis, and the input data typically concerns occurrence rates of basic events. Beyond the fault tree itself, the physical modelling in this part is generally confined to the determination of life distributions for components.

J. Flamm and T. Luisi (eds.), Reliability Data Collection and Analysis, 193–211.

The second part of a probabilistic risk assessment, the **accident consequence assessment,** concerns the consequences of an undesired event for man and his milieu. The type of data required for consequence assessment is more varied than for accident prediction, and there is no dominant methodology. Sophisticated physical models are used to describe the transport of hazardous substances through the biosphere, following a large release. Biological models describe the movement of hazardous substances through the food chain, and toxicological models predict the effect of exposures in the human population. In addition to quantifying parameters in these models, information may be required on weather conditions, ground water transport, emergency countermeasures, long term removal, etc, etc.

In both parts of risk assessment, expert judgment data has been used wherever objective data is lacking or unreliable. It is appropriate at this juncture to say a few words about the philosophical position taken in this document regarding the use of expert judgment as a source of data in scientific studies.

Expert judgment is by its nature uncertain, and using expert judgment involves choosing a methodology for representing experts' uncertainty. Many representations are under discussion in various fields. In this document it is assumed that uncertainty is represented as subjective probability. The reasons for this are extensively discussed elsewhere (Cooke, 1991) and will not be rehearsed here. Suffice to say that subjective probability provides the best mechanism for satisfying general principles for scientific methodology (for an extended discussion see Cooke, 1991):

Traceability: All data must in principle be traceable to its source, enabling a full scientific review by peers.

Neutrality: The method for processing expert judgment data should not bias experts to give assessments at variance with their true opinions.

Empirical control: Expert judgment should in principle be suceptible to empirical control.

Fairness: Experts' judgments should be treated equally, except as indicated via the mechanism of empirical control.

Both accident prediction and accident consequence assessment ultimately yield information regarding the occurrence rates of events. In accident prediction the input data is also of this form. As determinations of occurrence rates (either from data or via expert judgment) are generally uncertain, assessments of occurrence rates are commonly given in terms of a median value and a 90% central confidence band. Occurrence rates in consequence assessment, e.g. as expressed in complementary cumulative distribution functions for numbers of fatalities, represent the end of a long chain of reasoning. Because of modeling assumptions made along the way, input data, with attendant uncertainty, is required for occurrence rates (e.g. of weather conditions) *and* for modeling parameters. These latter can and should be expresesed in terms of occurrence rates of observable phenomenon, and this task requires insight into the models themselves. For the most part this document concerns expert assessments of the of occurrence rates of events. In discussing the question formulation phase, the issue of expressing uncertainty over modeling parameters is addressed.

This chapter is written to support the risk analyst who may have to use expert judgment data to quantify portions of a risk study. It is assumed that the analyst has access to and is familiar with the European Community software tool EXCALIBR (Cooke and Solomatine 1990) for processing expert judgment, or something equivalent. It is further assumed that he/she is familiar with the basic concepts of risk analysis, and that the analyst and experts are familiar with the respresentation of uncertainty as subjective probability.

1. PROBLEM IDENTIFICATION PHASE

Expert judgment will be applied for quantifying the occurrence rate of events satisfying one of the following characteristics:

- Historical data are not available

or

- Historical data are available but not sufficient for assessing the

occurrence rate.

The second condition will apply, for example, when historical data

- contains failure data but does not specify the reference class from which the data is drawn

and/or

- concerns events nominally similar to the event in question, yet does not specify relevant environmental or operational characteristics to be the same as those under which the events in question must be analysed.

For example, maintenance data may contain information on the number of failures on tests, but may not contain the number of tests; incident data banks may record the number of incidents but may not define the population from the incidents are drawn. Component reliability data banks frequently do not specify the operating characteristics under which their data has been gathered. In some applications this fact alone renders the data useless. In the field of aerospace, for example, the space environment is sufficiently unlike conditions on earth as to render data gathered on earth non-applicable for many components. Similarly, human error data drawn from simulator experiments may not apply to real situations in which factors like stress and confusion strongly influence behavior.

In deciding to apply expert judgment to assess the occurrence rate of a given event, it is incumbent upon the analyst to document the available historical data and indicate his reasons for not using this data. Even when existing historical data is not used, it is important to document the occurrence rates derivable/retrievable from this historical data. This will provide a type of check for the results of expert assessments. Large deviations must be explainable.

2. EXPERT IDENTIFICATION PHASE

The term "expert" is not defined by any quantitative measure of resident knowledge. Rather "expert for a given subject" is used here to designate a person whose present or past field contains the subject in question, and who is regarded by others as being one of the more knowledgeable about the subject. Such persons are sometimes designated in the literature as "domain" or "substantive" experts, to distinguish them from "normative experts" who are experts in statistics and subjective probability.

Identifying experts for a given subject therfore means identifying persons whose work terrain contains the subject and who are regarded as knowledgeable by others. Ideally, the analyst should consult *all* experts for a given subject. In practice this is generally impossible, and a selection of experts must be made. The method of selecting experts could significantly bias the results. It is therefore essential that experts be selected according to a traceable procedure which minimizes ad hoc decisions by the analyst. Two procedures are discussed below. The **Round Robin** procedure generates a superset of experts for a given subject, and the **Paired Comparison Voting Procedure** yields a prioritization of this superset.

The **Round Robin** and consists of the following steps:

1. Some names of potential experts are generated within the organization responsible for the study (if the organization could not do this, they could not perform the study in the first place). These persons are approached and asked:

- what is your background and knowledge base with regard to the subject?
- which other persons are knowledgeable with regard to the subject?

2. The persons named in the first round are approached with the same two questions.

3. Step 2 is iterated until (a) no new names appear, or (b) it is judged that a sufficiently diverse set of experts is obtained.

The knowledge base describes the type of information on which the experts' assessments would be based. This may be either

- articles in journals or technical reports
- experimental or observational data
- computer modeling.

The round robin provides a mechanism for generating names, but does not provide a mechanism for eliminating or prioritizing names. In this sense the procedure may be said to yield a superset of experts.

The **Paired Comparison Voting Procedure** prioritizes a set of experts for a selected issue. It is assumed that a superset of experts has been generated, perhaps by the round robin method described above. The superset is assumed to include experts themselves and also individuals who could identify most knowledgeable people for the question at hand. The paired comparison voting procedure enables this superset to prioritize itself. The procedure is as follows:

1. Each member of the superset completes a paired comparison exercies. The exercise consists of answering questions of the following type, for each pair of members of the set (including himself)

    Place an X *before the name of the person who seems most knowledgeable with regad to* <question at hand> (*void comparisons are allowed*).

    ——— *Expert # i*

    ——— *Expert # j*

2. The responses are processed using the paired comparisons software module included in EXCALIBR. The individual responses are analysed with regard to consistency and agreement.

3. The software tool generates a ranking.

The method of paired comparisons has proven to be a friendly method for building consensus with regard to a rank ordering of alternatives based on qualitative judgments. It has never been applied in the above manner, but would seem to recommend itself for the task of prioritizing experts. Judging individuals pairwise is cognitively much easier than choosing a "best expert" from the entire set. However, each member is required to judge all pairs of names; if there are n names, then there are n(n–1)/2 pairs. For n = 20 this entails 190 comparisons for every member. In practice, for n greater than 6 or 7, restricted sampling techniques will have to be used.

Important: there is no methodological reason to prioritize the set of experts. Prioritization must not replace performance - based scoring. It is recommended only if budgetary constraints limit the number of experts and scrutibility requirments preclude informal selection procedures.

3. EXPERT CHOICE PHASE

After the set of experts is identified, a choice is made which experts to use in the study. In general, the largest number of experts consistent with the level of available resources should be used. In any event at least four experts for a given subject should be chosen. The choice should made so as to diverisfy the knowledge bases and institutions of employment.

The choice of experts is one of the most sensitive aspects of the expert judgment process. If this is not performed in a fair and objective manner, the credibility of the results will surely be compromised. On the other hand, the classical model for combining expert judgments insures that it is *impossible* to predetermine the results by appropriately selecting experts. Indeed, the classical model forms a weighted combination of expert judgments according to a scoring rule optimization routine. Several features of this routine serve to robustify the results against the choice of experts:

- the procedure is fully traceable and objective
- an expert's weight is deteremined via classical hypothesis testing
- the number of experts is not limited by computational constraints
- the model is generally insensitive to "swamping"...increasing the numerical representation of a "point of view" in the expert panel does not directly affect the results.

This having been said, the objectivity of the results could of course be compromised by an expert or analyst with a strong motivaltional bias and sufficient insight into the model. If there is reason to be apprehensive in this regard, then expert judgment should not be used.

4. QUESTION FORMULATION PHASE

4.1 EVENT DEFINITION

In defining the events about which experts will be asked, two golden rules apply:

1. Ask only for values of observable or potentially observable quantities.

2. Formulate questions in a manner consistent with the way in which an expert represents the relevant information in his knowledge base.

The first rule follows from the fact that subjective probability represents uncertainty with regard to potentially observable phenomena. The second rule typically constrains the choice of dimensions and units.

Probabilistic risk analysis requires probabilities of occurrence for basic events in a fault tree. It is customary to express uncertainty by stating 5% and 95 % confidence bands for these probabilities. However, this way of speaking invites confusion. If the probabilities of occurrence are *subjective* probabilities, then an additional expression of subjective uncertainty makes no sense. Indeed, subjective probabilities represent uncertainty with regard to observable quantities; a subjective probability is not itself an observable quantitiy and hence it is not meaningful to assign a subjective probability to a subjective probability[1].

The questions put to the expert should of yield answers in the form of (quantiles of) subjective probability distributions over potentially observable quantities. A "probability" is not directly observable, but a limiting relative frequency of occurrence is. Therefore, one should not ask "what is the probabilty of event X", but "what is the limiting relative frequency of X" (with the reference class appropriately specified).

The events thus queried must be repeatable. It would, in principle, be possible to ask for subjective probabilities of occurrence for unique, i.e. non-repeatable, events. For example, "What is the probability of event X next year at this particular installation?" Rather than (quantiles of) subjective distributions, this type of questioning yield answers in the form of discrete probability judgments: "the probability of the X at this installation next year is p". This approach is intrinsically sound, and widespread in the field of meteorological forecasting; however, it encounters a number of practical difficulties in risk analysis:

---

[1] This point has been amply discussed in the early theoretical literature, but continues to cause confusion among practitioners. Another way to appreciate this point is the following: Theoretically speaking, a subjective probability for the occurrence of event X is a betting rate, e.g. a willingness to wager on the occurrence of X. An individual can be unsure of his own betting rate in the sense that he might change this rate very quickly given new reflection or evidence. However, he cannot be uncertain of this rate in the sense of uncertainty represented by subjective probability. Indeed, he cannot meaningfully wager that his betting rate for X will fall within some non-trivial interval, as the betting rate for X is something which he decides for himself. This is why we say that subjective probability represents uncertainty with regard to observable phenomena.

- it does not represent uncertainty in terms of 5%, 50% and 95% quantiles for objective occurrence rates, as is customary in risk analysis
- discrete probability assessments are more difficult to process and require a much larger number of calibration variables
- assessments of unique events are not transportable outside their original context, whereas the analyst will frequently want to apply assessments from previous studies.

These rules will be illustrated with an example. Suppose the analyst is interested in the event that a manned spacecraft is penetrated by a particle of space debris during a future space mission. An expert on space debris may be able to assess the occurrence rate of impacts by particles with diameter and velocity exceeding given values, on a randomly oriented surface per square meter, per time unit, in a given orbit. He/she will not know the diameter and velocity values necessary to penetrate a given spacecraft[2]. An expert on shielding could give an informed opinion on the critical values for penetration. Neither of these experts will know the spacecraft surface area or the length of time which the spacecraft will be used.

Hence, no expert will be directly asked for the occurrence rate of spacecraft penetrations by debris particles. This quantity will be computed by the analyst on the basis of information supplied by experts. The analyst must know "who knows what", and must break his question down into parts which can be intellgently posed to the experts. Pursuing the above example for particle diameter, space debris experts will be asked for the occurrence rate of impacts per square meter (randomly oriented) per time unit of particles with diameter exceeding one or more given critical values.

The process of choosing the relevant occurrence events and specifying the units with respect to which limiting relative frequencies will be asked is called **event definition,** and results in a clauses of the form:

<occurrence event> **per** <units>

Filling the above example we get something like:

> *Impact with a debris particle with diameter greater than 1 cm on a randomly oriented surface per square meter per year*

---

[2] For the particles of interest, density is generally assumed constant. Shielding models express "critical thickness" of a shielding surface as functions of mass, density and velocity of the impacting particle, and of material constants characteristic of the shielding. In space stations, survivability is principally determined by the depressurization rate, and this in turn is determined by the diameter of a penetrating particle. The velocity and diameter distributions are generally considered independent, and the former is known empirically. For simplicity, the ensuing discussion is restricted to particle diameter.

The occurrence event should be (potentially) observable, and the units, e.g. "per mission", "per demand", "per year", "per cycle", etc., should be consistent with the experts' knowledge base representation, whenever possible. Experts are asked to assess their subjective probability distributions over the limiting relative frequencies of occurrence events, per unit.

4.2 EVENT CONDITIONALIZATION

When an occurrence event is defined, the analytst should determine those variables which influence (either positively or negatively) the event's rate of occurrence. The set of such variables is called the **causation set** for the event in question. In the example above, the impact event is influenced by

- the orbit
- date of the flight
- the growth rate of debris particles
- the time spent in orbit
- the surface area and orientation of the spacecraft

The growth rate of debris particles in turn is influenced by

- drag / solar cycle
- the launch rate
- the emergence of new nations in the space community
- military activity
- international agreements.

Isolating the causation set may require consultation with experts. After determining the causation set for the occurrence event in question, the analyst must identify those variables, if any, in the causation set whose values may be assumed fixed for the purposes of his study. For example, the study may presuppose a fixed orbit, a given date, a given shielding configuration, and no significant military activity affecting the debris flux.

The set of variables in the causation set whose values are fixed within the scope of the study is termed the **conditioning set**. Note that the conditioning variables mentioned above are *decision variables* for the design and operation of spacecraft. The designer and flight control center choose the orbit, the date, and would certainly not launch if significant military activity in space had taken place. The conditioning set frequently, but not always, consists of decision variables.

Other variables in the causation set, i.e. those whose values are *not* assumed fixed, belong to the **uncertainty set**. Uncertainty over the values of variables in the uncertainty set will contribute to uncertainty with regard to the occurrence event. Hence the causation set is decomposed into the conditioning set and the uncertainty set:

causation set = conditioning set $\cup$ uncertainty set

It is essential to check that the conditioning and uncertainty sets for the variables in the study are mutually consistent, and consistent with the presuppositions of the study as a whole. It may be necessary to redefine the scope of the whole study before proceding further.

Having determined the values of variables in the conditioning set, the analyst formulates the event with conditioning clause:

<occurrence event> **per** <units>

given <values of variables in conditioning set>

taking into account <variables in the uncertainty set>

For the space debris example, this might look like

Impact with a debris particle with diameter greater than 1 *cm on a randomly oriented surface* per *square meter* per *year*

given:
- *orbit* 500 *km*, *inclination* 30°
- *launch in the late* 1990*'s*
- *no significant military activity*

taking into account:
- *launch rate, international agreements unknown*
- *growth rate of debris particles uncertain*

4.3 QUESTION FORMULATION

The formulation of the question to the expert must indicate clearly which variables in the causation set are fixed, and at which values; and also which variables are not fixed and may contribute to the uncertainty of the occurrence event. The format proposed is:

How often will <occurrence event> *occur per* <units> *given* <values of variables in conditioning set> *taking into account* <variables in the uncertainty set>?

Please give your median assessment and your 90¢ *central confidence band for the limiting relative frequency.*

Your subjective probability is 50% that the true limiging relative frequency lies below your median assessment. Your subjective probability is 10% that the true limiting relative frequency falls outside your central confidence band (5% above and 5% below).

4.4 ELICITATION FOR CONSEQUENCE ASSESSMENT

Using expert judgment in accident consequence assessment can involve specific problems not encountered in accident prediction. These arise from the fact that in consequence assessment we are interested uncertainty with regard to functional relationships between observable variables. The use of such mathematical models introduces three types of uncertainty:

1. Model uncertainty: The models may be, and strictly speaking usually are, wrong. In most cases the models are derived under simplifying physical assumptions. At best these are taken to represent reasonable approximations.

2. Measurement uncertainty: The models may contain physical parameters which cannot be precisely measured.

3. Numerical uncertainty: The routines for solving the model may introduce simplifications and numerical errors.

We shall not address the subject of eliciting uncertain functional relationships, but refer the interested reader to Cooke and Vogt (1990) where this problem is treated in detail. Suffice to say that the such elicitations will result in conditional subjective distributions for some dependent variable Y, given values of some independent variables $X_1,...X_n$. For an arbitrary model $Y = M(X_1,...X_n)$, the above reference describes methods for fitting distributions to the parameters of M, given the conditional distributions. Y need not be a limiting occurrence rate, but can be any continuous or discretely distributed variable.

5. THE CALIBRATION VARIABLE SELECTION PHASE

The classical model for combining expert judgment data requires that experts assess their subjective probability distributions for calibration variables, i.e. variables whose values are or will become known to the analyst within the time frame of the study. Calibration variables are important not only for determining the weights for combining expert judgments. They provide for assessing the performance of the combined assessment (the "optimized decision maker"), and also form an important part of the feedback to experts, helping them to guage their subjective sense of uncertainty against quantative measures of performance. Two types of calibration variables are distinguised, namely those generated by the analyst himself, and those provided by supporting software. The latter are still at a prill stage of development, and the discussion is therefore confined to the former.

5.1 CALIBRATION VARIABLES GENERATED BY THE ANALYST

It is impossible to give an effective procedure for generating meaningful calibration variables. Here the creativity and resourcefulness of the analyst come into play. General guidelines and tips will be provided here.

Calibration variables falling squarely within the experts' field of expertise are called **domain variables.** In addition to domain variables, it is permissible to use variables from fields which are adjacent to the experts' proper field. These are called **adjacent variables.** Adjacent variables are those about which the expert should be able to give an educated guess. It will often arise that a given calibration variable is a domain variable for one expert and an adjacent variable for another expert. In the literature one sometimes encounters the use of **general knowledge** variables in an expert judgment context ("what is the population of Wroclaw, Poland"). For such variables an educated guess from the expert could not be expected, and the use of general knowledge variables is not recommended.

Calibration variables may also be distinguished according to whether they concern **predictions** or **retrodictions.** For predictions the true value does not exist at the time the question is answered, whereas for retrodictions, the true value exists at the time the question is answered, but is not known to the expert.

The following examples taken from expert judgment exercises in the aerospace field illustrate the above terminology.

> *percentage successful launches with rocket type X from* 1960 *up to the present*
> (domain, retrodiction)
>
> *observed frequency of airplane crashes within* 1 *km of an aircraft carrier per* km^2 *per airplane overflight*
> (adjacent, retrodiction)
>
> *number of tracked debris particles injected into orbit in* 1985
> (domain, retrodiction)

number of tracked debris particles injected into orbit in 1990
(domain precidtion)

In general, domain predictions are the most meaningful, in terms of proximity to the items of interest, and are also the hardest to generate. Adjacent retrodictions are easier to generate, but are less closely related to the items of interest. The use of adjacent retrodictions is sanctioned by the supposition that performance on such variables correlates with performance on the items of interest. There is no direct proof of this supposition at present, but it has been found that "experience" positively correlates with performance on adjacent retrodictions, and does not correlate with performance on general knowledge variables (see Cooke et al. "Calibration and information in Expert resolution; a classical approach" **Automatica**, vol. 24 no. 1 pp 87–94, 1988).

6. THE ELICITATION PHASE

Elicitation should be performed with each expert individually. If possible, the analyst should be present, and the session should not greatly exceed one hour.

6.1 FORMAT

The elicitation must provide answers to the questions formulated according to paragraph **5**. It is recommended to strive for a uniform, actractive format that allows both graph and numerical responses. In assessing rates of occurrence which are much smaller than one in the relevant unit, the following format in semi–log scale has been found statisfactory.

6.2 CROSS CHECKS

Three types of cross check can be performed in the elicitation of each variable:

1) The analyst should ask the expert if he has considered all possible values of variables in the uncertainty set, in determining his assessment.

event occurrence question (see 6.3)

Please place an X at your median assessment and draw an interval indicating your 90% central confidence band

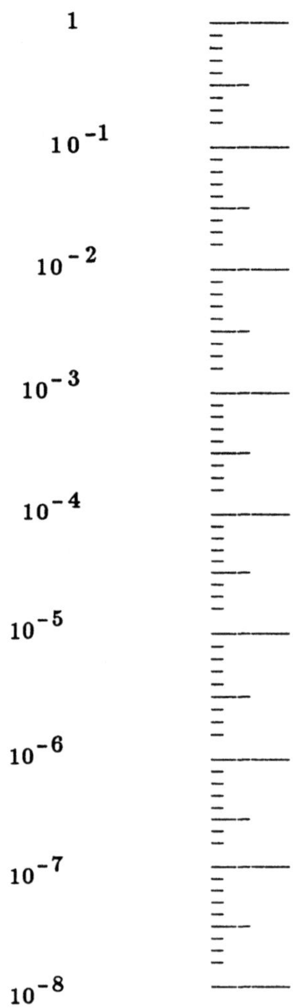

2) The expert's uncertainty assessment is cross checked by asking the expert to fill in the "degree of surprise" format shown below:

How often would you expect the true value of <occurrence event> to be greater than your median assessment by at least	
a factor 2	a factor 10
one in three	one in three
one in five	one in five
one in ten	one in ten
one in twenty	one in twenty
one in fifty	one in fifty
one in 100	one in 100
more than one in 100	more than 100

The upper confidence bound should be associated with a degree of surprise of one in twenty. If the ratio (95% quantile)/(50% quantile) is less than 2, then the answer in the first column should not be more frequent than "one in twenty"; if this ratio is between 2 and 10, then the answers in the first and third columns should straddle "one in twenty"; if this ratio is greater than 10, then the answer in the third column should be more frequent than "one in twenty".

3) Cross checks on the coherence of the experts' determine whether his assessments are consistent with the axioms of probability. Such checks can be performed in a number of ways. For example, if A and B are disjoint rare events, the expert could be asked about A and B individually, and also about "A or B". The latter assessment should be roughly the sum of the previous two assessments. A word of caution is appropriate here. It is clear that assessments will never be perfectly coherent. A system is decomposed into basic events for which the analyst believes that his experts can give the best assessments. Coherent assessments for all other events can only be obtained by very complicated computations. For example, an assessment for the top event of a system could only be consistent with assessments for the basic events if the expert has solved the minimal cutset equation – a formidible task even for

small systems. Moreover, the assessments concern *distributions* of failure frequencies, and it is not in general possible to compute the *distribution* of failure frequencies for the union of two events from the individual frequency distributions. For example, the median occurrence probability of the event "A or B" is not the sum of the median occurrence probabilities for A and B. In checking coherence, the analyst should:

-be coulant

-avoid issuing corrections

-attend only to gross violations of the probability axioms.

Cross checks are tools to help the expert express his uncertainty. As soon as the expert feels comfortable with the assessment task, cross checks may become annoying and should be omitted.

6.3 DRY RUN

After completion of the definition and formulation phase, it is strongly advised to perform a dry run before finalizing the elicitation formats. This means running through the elicitation process with one selected expert. This enables the analyst to identify unnecessary ambiguities in formulation, and generally bebug the whole process. In particular, the analyst should verify that the causation set for each occurrence event is complete.

7. THE COMBINATION PHASE

Various models for combining expert judgments are described in (Cooke, 1991). Of these, only the simple classical model is recommended for implementation at the present. This model is detailed in chapter 12 of that reference.

The classical model requires software support, such as provided in the tool EXCALIBR. The user's manual should be consulted for details on the program.

8. THE DISCREPANCY/ROBUSTNESS ANALYSIS AND FEEDBACK PHASE

After the elicitation process, two types of post hoc activities should be undertaken, namely discrepancy/robustness analysis and feedback to the experts.

8.1 DISCREPANCY/ROBUSTNESS ANALYSIS

After determining the optimal decision maker the analyst should perform discrepancy and robustness analyses. The robustness analysis should determine the sensitivity of the results to the choice of calibration variables. This is accomplished in the following steps:

1. Remove one calibration variable from the calibration set.

2.Repeat the combination process, determining a new optimized decision maker's distributions for the variables of interest.

3. Compare the new decision maker's distributions with those of the original optimized decision maker.

4. Repeat steps 1 – 3 for each calibration variable.

Step 2 is quite easily performed, using the filtering feature of the tool EXCALIBR. Step 3 is not software supported at present. Experience has shown the that classical model is reasonably robust with respect to the above procedure. In general the virtual weights of the optimized decision maker and the weights of the individual experts are more sensitive to the deletion of calibration variables than the decision maker's distributions for the variables of interest.

Discrepancy analysis is accomplished in the following steps:

1. Scan the experts' assessments for the varibles of interest and select those variables for which the experts' disagreement is unusually sharp.
2. Check whether the causation sets for these variables is really complete. It may be that some experts tacitly regard other variables as causally significant, and different assumptions regarding the values of these variables is responsible for the disagreement.
3. If step 2 leads to the identification of new causally significant variables, and if the matter is of sufficient importance, reformulate the the questions and repeat the elicitation.

Step 1 is not software supported at present.

Discrepancy analysis could in principle be performed on the calibration variables, as well as the variables of interest. In practice, however, this is generally not feasible, as it will usually be possible for the expert to retrieve the true values of the calibration variables, and this would make re–elicitation pointless.

Discrepancy analysis should also take account of the calibration of the decision maker on the calibration variables. High decision maker calibration generates confidence in the results of the analysis. Low decision maker calibration on the other hand forces the analyst to address the question whether his expert data need not be developed further. Further development might include:

– expanding the set of experts

– re–elicitation of expert judgment following an expert training in probabilistic assessment.

8.2 FEEDBACK

Each expert must have access to

- his/her assessments
- his/her calibration and entropy scores
- his/her weighing factors
- passages in which his/her name is used.

In addition they should receive enough information to interprete the information proffered both in general terms and in relation to their performance on the calibration variables. This may help them in neutralizing common subjective biases. In particular, the expert judgment combination software could output for each expert

- whether the expert shows a tendency toward over– or underconfidence
- whether the expert shows a tendency to over– or underestimate.

9. THE DOCUMENTATION AND COMMUNICATION PHASE

In conducting a full–scale expert judgment study it is desirable to break the documentation down into two phases, denoted here **Preliminary** and **Final documentation**. The formats proposed below should be regarded as provisional suggestions.

9.1 PRELIMINARY DOCUMENTATION

Preliminary documentation serves to enable a review of all decisions that have been taken prior to the actual acquisition of the expert judgment data. The following topics should be covered:

Motivation: What are the objectives of the study, and why is it necessary to use expert judgment?

Survey of expertise: Who are the experts, what is their knowledge base and affiliation.

Selection of experts: Which experts will participate in the study and how were they selected?

9.2 FINAL DOCUMENTATION

The final documentation should contain all information to enable a full scientific review. This entails that the following points be addressed:

Preliminary documentation update: Where necessary, changes in motivation, survey of expertise or selection of experts are documented.

Elicitation information: Copies of elicitation formats, descriptions of elicitation procedure, formulations of questions, motivation of choice of calibration variables.

Expert judgment data: All responses from each expert (suitably coded), calibration and information, weighting factors from each expert and from the optimized decision maker.

Discrepancy/Robustness analysis: Description of the activities described in section 8.1.

ACKNOWLEDGEMENT
This work was carried out at the Delft University of Technology under contract with the European Community, and was initially reported in Cooke and Solomatine (1990). The procedures described in this chapter are closely related to procedures developed under contract with the European Space Agency. These in turn have drawn on experiences and insights from many people and organizations who have carried out expert judgment studies. Institutions which have supported these applications are gratefully acknowledged:

The Dutch Ministery of Environment
Shell
DSM
Delft Hydraulics Laboratory
The European Community
The National Radiation Protection Board
Kernforschungszentrum Karlsruhe
The European Space Agency

REFERENCES

Cooke, R. (1991) **Experts in Uncertainty,** Oxford University Press

Cooke R. and Vogt, F (1990) "Parameter Fitting for Uncertain Models" included as Annex I to **The European Communities' Expert Judgment Study on Atmospheric Dispersion and Deposition**,(R. Cooke), Department of Mathematics, Delft University of Technology, August 1991.

Cooke R. and Solomatine, D. (1990)"EXCALIBR software package for expert data evaluation and fusion in risk and reliability assessment", Department of Mathematics, Delft University of Technology, June 1990.

ON THE COMBINATION OF EVIDENCE IN VARIOUS MATHEMATICAL FRAMEWORKS

Didier DUBOIS and Henri PRADE
Institut de Recherche en Informatique de Toulouse
Université Paul Sabatier, 118 route de Narbonne
31062 TOULOUSE Cedex - FRANCE

1. Introduction

The problem of combining pieces of evidence issued from several sources of information turns out to be a very important issue in artificial intelligence. It is encountered in expert systems when several production rules conclude on the value of the same variable, but also in robotics when information coming from different sensors is to be aggregated. Solutions proposed in the literature so far have often been unsatisfactory because relying on a single theory of uncertainty, a unique mode of combination, or the absence of analysis of the reasons for uncertainty. Besides dependencies and redundancies between sources must be dealt with especially in knowledge bases, where sources correspond to production rules.

In this paper, we present a conceptual framework for the combination of evidence issued from various sources. We particularly emphasize the case of *parallel* sources as opposed to the problem of belief updating where sources do not play a symmetric role. A typology of encountered situations is proposed, ranging from the case when *all* sources are reliable to the case when only very weak knowledge about their reliability is available ; especially it is not always realistic to assume that the reliability of sources is precisely known. For instance it may be assumed that "one of the sources is reliable" but without knowing which one it is. The proposed approach to combination is general enough to tackle various situations, but also to be implemented within various mathematical frameworks such as possibility theory, probability theory, and Shafer belief functions. It is easily expressed in generalized set-theoretic terms. Especially conjunctive combinations (based on set-intersection) apply when sources are reliable, while disjunctive combinations (based on set-union) deal with the case of a reliable source hidden in a group of other sources. Most combination operations lie in between these two extreme modes. They assume less information than what a Bayesian approach to combination requires (especially no a priori information), but the quality of the result may be poorer.

J. Flamm and T. Luisi (eds.), Reliability Data Collection and Analysis, 213–241.

First a short background on the main existing frameworks for the representation of uncertainty is provided. Then we consider the general problem of combining uncertain pieces of information obtained from different sources (which are supposed to play a symmetric role). The paper also gives a formulation of the various combination modes in several mathematical settings thus unifying several rules proposed independently by several authors, e.g. MYCIN rule, Dempster rule, fuzzy set operations, etc.... These rules are reviewed and classified according to the situations where they can be used and the available information which is assumed. Then the belief updating problem, where one source play the role of a priori information is analyzed, when the updating is due to several concurrent sources. Lastly, a section addresses various questions, especially how to deal with the possible existence of a (partial) conflict between the sources, the reliability of the sources, the nature of the available information : generic or specific, the dependencies between sources, ...

2. Uncertainty as Imprecise Probability

Let $\mathcal{P}$ be a Boolean algebra of propositions, denoted by **a**, **b**, **c**... ; **0** and **1** will denote the contradiction and the tautology respectively (i.e. the least and the greatest elements in $\mathcal{P}$). Let P(**a**) denote the probability attached to proposition **a**. Then P(**a**) = 1 means that **a** is certainly true, while P(**a**) = 0 means that **a** is certainly false. In this section is recalled a framework that unifies various proposals for representing uncertainty and formally relates them to probability theory (Dubois & Prade [14], [15]).

2.1. Upper and Lower Probabilities

In situation of partial ignorance, the probability of **a** is only imprecisely known, and can be expressed as an interval range [C(**a**), Pl(**a**)] whose lower bound can be viewed as a degree of certainty (or belief) of **a**, while the upper bound represents a grade of plausibility (or possibility) of **a**, i.e. the extent to which **a** cannot be denied. Total ignorance about **a** is observed when there is a total lack of certainty (C(**a**) = 0) and complete possibility (Pl(**a**) = 1) for **a**. A natural assumption is to admit that the evidence which supports **a** also denies 'not **a**' (¬**a**). This modeling assumption leads to the convention

$$C(\mathbf{a}) = 1 - Pl(\neg \mathbf{a}) \qquad (1)$$

which is in agreement with P(**a**) = 1 - P(¬**a**), where P(**a**) and P(¬**a**) are imprecisely known probabilities. This equality also means that the certainty of **a** is equivalent to the impossibility of ¬**a**. The framework of probability theory does not allow for modelling the difference between possibility and certainty, as expressed by (1). Functions C and Pl are usually called lower and upper probabilities, when considered as bounds on an unknown probability measure. See

Walley & Fine [44] and Dubois & Prade [15] for surveys on upper and lower probabilities.
Using (1), the knowledge of the certainty function C over the Boolean algebra of propositions $\mathcal{P}$ is enough to reconstruct the plausibility function Pl. Especially the amount of uncertainty pervading **a** is summarized by the two numbers C(**a**) and C(¬**a**). They are such that C(**a**) + C(¬**a**) ≤ 1, due to (1). The above discussion leads to the following conventions, for interpreting the number C(**a**) attached to **a** :

i) C(**a**) = 1 means that **a** is certainly true.
ii) C(**a**) = C(¬**a**) = 0 (i.e. Pl(**a**) = 1) means total ignorance about **a**. In other words **a** is neither supported nor denied by any piece of available evidence. This is a self-consistent, absolute reference point for expressing ignorance.
iii) C(**a**) = C(¬**a**) = 0.5 (i.e. Pl(**a**) = 0.5) means maximal probabilistic uncertainty about **a**. In other words the available evidence can be shared in two equal parts : one which supports **a** and the other which denies it.This is the case of pure randomness in the occurence of **a** .
iv) C(¬**a**) = 1, i.e. Pl(**a**) = 0 means that **a** is certainly false.

Note that total ignorance implies that we are equally uncertain about the truth of **a** and ¬**a**, as well as when C(**a**) = C(¬**a**) = .5. In other words ignorance implies uncertainty about the truth of **a**, but the converse is not true. Namely, in the probabilistic case, we have a lot of information, but we are still completely uncertain. Total uncertainty is more generally observed whenever C(**a**) = C(¬**a**) ∈ [0,0.5]. The amount of ignorance is assessed by 1 - 2 C(**a**) in that case.

2.2. Plausibility and Belief Functions

The mathematical properties of C depend upon the way the available evidence is modelled and related to the certainty function. In Shafer theory [37], a body of evidence ($\mathcal{F}$,m) is composed of a subset $\mathcal{F} \subseteq \mathcal{P}$ of n focal propositions, each being attached a relative weight of confidence m($\mathbf{a}_i$) for all $\mathbf{a}_i \in \mathcal{F}$. m($\mathbf{a}_i$) is a positive number in the unit interval ; m is called a basic probability assignment and satisfies the following constraints

$$\Sigma_{i=1,n}\ m(\mathbf{a}_i) = 1 \qquad (2)$$

$$m(\mathbf{0}) = 0 \qquad (3)$$

The requirement (3) expresses the fact that no confidence is committed to the contradictory proposition. The weight m(**1**) possibly granted to the tautology represents the amount of total ignorance since the tautology does not support nor deny any other proposition. The fact that a proposition **a** supports another proposition **b** is formally expressed by the logical entailment, i.e. $\mathbf{a} \to \mathbf{b}$ (= ¬**a** ∨ **b**) = **1**. Let S(**a**) be the set of propositions supporting **a** other than the contradiction **0**. The function C(**a**) is called a *belief function* in the sense of Shafer (and denoted 'Bel') if and only if there is a body of evidence ($\mathcal{F}$,m) such that

$$\forall \mathbf{a},\ \mathrm{Bel}(\mathbf{a}) = \Sigma_{\mathbf{a}_i \in S(\mathbf{a})}\, m(\mathbf{a}_i) \qquad (4)$$

$$\forall \mathbf{a},\ \mathrm{Pl}(\mathbf{a}) = \Sigma_{\mathbf{a}_i \in S(\neg \mathbf{a})^c - \{\mathbf{0}\}}\, m(\mathbf{a}_i) \qquad (5)$$

where 'c' denotes complementation. Clearly, when the focal elements are only atoms of the Boolean algebra $\mathcal{P}$ (i.e. $S(\mathbf{a}_i) = \{\mathbf{a}_i\}$, for all i = 1,n) then $\forall \mathbf{a}$, $S(\neg\mathbf{a}) = S(\mathbf{a})^c - \{\mathbf{0}\}$, and $\mathrm{Pl}(\mathbf{a}) = \mathrm{Bel}(\mathbf{a})$, $\forall \mathbf{a}$. We recover a probability measure on $\mathcal{P}$. In the general case the quantity $\mathrm{Pl}(\mathbf{a}) - \mathrm{Bel}(\mathbf{a})$ represents the amount of imprecision about the probability of **a**. Interpreting the Boolean algebra $\mathcal{P}$ as a family of subsets of a referential set Ω of possible worlds, the atoms of $\mathcal{P}$ can be viewed as forming a partition of Ω. Then a focal proposition $\mathbf{a}_i$ whose model is the subset $M(\mathbf{a}_i) = A_i \subseteq \Omega$ corresponds to the statement : there is a probability $m(\mathbf{a}_i)$ that the information about the location of the actual world can be described by A_i. When A_i is not a singleton, this piece of information is said to be imprecise, because the actual world can be anywhere within A_i. When A_i is a singleton, the piece of information is said to be precise. Clearly, Bel = Pl is a probability measure if and only if the available evidence is precise (but generally scattered between several disjoint focal elements viewed as singletons).

Note that although $\mathrm{Bel}(\mathbf{a})$ and $\mathrm{Pl}(\mathbf{a})$ are respectively lower and upper probabilities, the converse is not true, that is any interval-valued probability cannot be interpreted as a pair of belief and plausibility functions in the sense of (4) and (5). Indeed for any function C from a finite Boolean algebra $\mathcal{P}$ to [0,1], there is another real-valued function m with domain $\mathcal{P}$ such that (4) holds for Bel = C. This is called Moebius inversion (see [37]). However the fact that m is a <u>positive</u> function, i.e. $\forall\, \mathbf{a} \in \mathcal{P}$, $m(\mathbf{a}) \geq 0$ is a characteristic feature of belief functions.

2.3. Possibility Measures

When two propositions **a** and **b** are such that $\mathbf{a} \in S(\mathbf{b})$, we write $\mathbf{a} \vdash \mathbf{b}$, and $\vdash$ is called the entailment relation. Note that $\vdash$ is reflexive and transitive and equips the set $\mathcal{F}$ of focal elements with a partial ordering structure. When $\mathcal{F}$ is linearly ordered by $\vdash$, i.e., $\mathcal{F} = \{\mathbf{a}_1, \ldots, \mathbf{a}_n\}$ where $\mathbf{a}_i \vdash \mathbf{a}_{i+1}$, i = 1,n - 1, the belief and plausibility functions Bel and Pl satisfy the following properties [37]

$$\mathrm{Bel}(\mathbf{a} \wedge \mathbf{b}) = \min(\mathrm{Bel}(\mathbf{a}), \mathrm{Bel}(\mathbf{b})) \qquad (6)$$

$$\mathrm{Pl}(\mathbf{a} \vee \mathbf{b}) = \max(\mathrm{Pl}(\mathbf{a}), \mathrm{Pl}(\mathbf{b})) \qquad (7)$$

Formally, the plausibility function is then a possibility measure in the sense of Zadeh [52]. The following equivalent properties due to the duality (1), then hold

$$\begin{cases} \max(\mathrm{Pl}(\mathbf{a}), \mathrm{Pl}(\neg\mathbf{a})) = 1 \\ \min(\mathrm{Bel}(\mathbf{a}), \mathrm{Bel}(\neg\mathbf{a})) = 0 \\ \mathrm{Bel}(\mathbf{a}) > 0 \Rightarrow \mathrm{Pl}(\mathbf{a}) = 1 \end{cases} \tag{8}$$

In the following possibility measures are denoted Π for the sake of clarity. The dual measure through (1) is then denoted N and called a necessity measure [9]. Zadeh [52] introduces possibility measures from so-called possibility distributions, which are mappings from Ω to [0,1], denoted π. A possibility and the dual necessity measure are then obtained as

$$\forall A \subseteq \Omega,\ \Pi(A) = \sup\{\pi(\omega) \mid \omega \in A\} \tag{9}$$

$$\forall A \subseteq \Omega,\ N(A) = \inf\{1 - \pi(\omega) \mid \omega \in A^c\} \tag{10}$$

and we then have $\pi(\omega) = \Pi(\{\omega\})$, $\forall\omega$. The function π can be viewed as a generalized characteristic function, i.e. the membership function μ_F of a fuzzy set F [51]). Let F_α be the α-cut of F. i.e., the subset $\{\omega \mid \mu_F(\omega) \geq \alpha\}$ with $\pi = \mu_F$. It is easy to check that in the finite case, the set of α-cuts $\{F_\alpha \mid \alpha \in (0,1]\}$ is the set $\mathcal{F}$ of focal elements of the possibility measure Π. Moreover, let $\pi_1 = 1 > \pi_2 \ldots > \pi_n$ be the set of distinct values of $\pi(\omega)$, let $\pi_{n+1} = 0$ by convention, and A_i be the π_i-cut of F, i = 1,n. The basic probability assignment m underlying Π is completely defined in terms of the possibility distribution π as [8] :

$$\begin{cases} m(A_i) = \pi_i - \pi_{i+1} \quad i = 1,n \\ m(A) = 0 \text{ otherwise} \end{cases} \tag{11}$$

Figure 1 gives an illustration of relation (11) between $\pi = \mu_F$ and m.

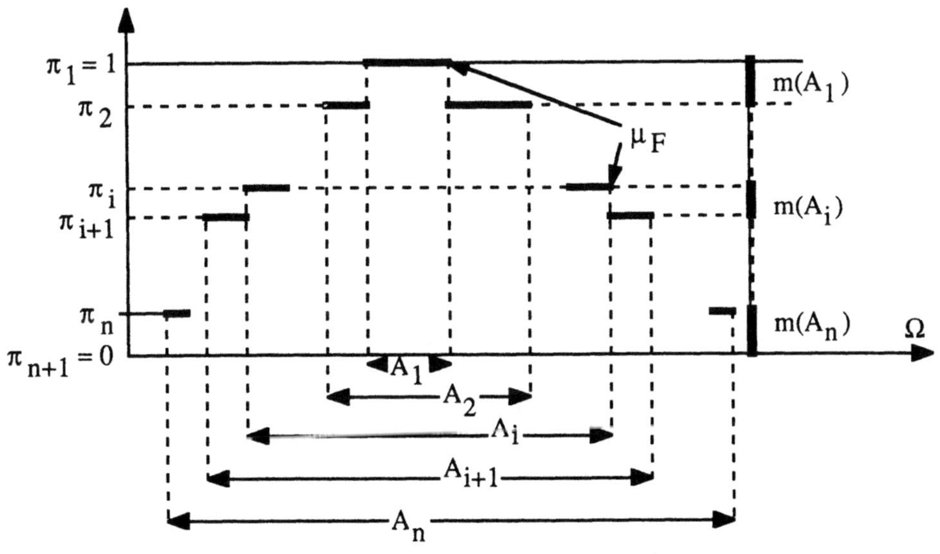

Figure 1 : Nested focal elements giving birth to a fuzzy set

Interpreting N(**a**) as a belief degree as in MYCIN and N(¬**a**) as a degree of disbelief in **a**, (6), (7) and (8) are assumed by Buchanan & Shortliffe [2] to be valid. Hence, as noted by Prade [33], MYCIN's treatment of uncertainty is partly consistent with possibility theory.
Nguyen [31] pointed out that the basic probability assignment m underlying a belief or a plausibility function defines a non-empty random subset of Ω ; indeed m is a probability assignment over the power set 2^{Ω}. Thus mathematically speaking belief functions can be viewed equivalently in terms of random sets. Especially the set of belief functions on Ω includes the power set 2^{Ω}. To see it, it is enough to notice that a subset $A \subseteq \Omega$ is equivalent to the belief function based on the body of evidence ($\mathcal{F}$,m) such that $\mathcal{F} = \{A\}$ and m(A) = 1. Moreover possibility measures correspond to fuzzy sets. These remarks stress the fact that belief functions are generalized sets as well as generalized probability measures. The set- theoretic point of view discussed at length elsewhere [12] was not present in Shafer's book [37].
Two recently proposed frameworks for the representation of uncertainty have been introduced here as particular case of upper and lower probability systems. It should be clear that they can also be viewed as *distorted* probability systems, i.e. as based on axioms which are distorted versions of the additivity axiom of probabilities ; in that view upper and lower probabilities are no longer considered as bounds on ill-known probability values but are numerical, precise translation of subjective degrees of plausibility and certainty. See Dubois & Prade [15] for instance for a parallel presentation of the upper and lower probability view and of the distorted probability view.

Remark : For simplicity, we have presented plausibility and belief functions, as well as possibility and necessity measures in a finite setting. The above definitions can be extended to non-finite spaces like $\mathbb{R}^n$ (where $\mathbb{R}$ denotes the real line) ; this is almost straightforward with possibility and necessity measures.

2.4. Normalization and Discounting

Assume a piece of uncertain information is represented by means of a basic probability assignment m on 2^{Ω} (i.e. an assignment satisfying (2)). The constraint (3), i.e. in terms of subsets m(Ø) = 0, expresses that we are certain that the random set defined by m is not empty, or in other words that the actual state of the world exists and is somewhere in Ω. When m(Ø) = 0, m is said to be *normal*. In case of a possibility measure, m is equivalent to a possibility distribution π as already said. The condition m(Ø) = 0 guarantees that $\pi_1 = 1$ in (11), i.e. $\exists\, \omega \in \Omega$, $\mu_F(\omega) = \pi(\omega) = 1$; the fuzzy set F is then said to be *normal*. More generally, the height of a fuzzy set F defined by $h(\mu_F) = \sup_{\omega \in \Omega} \mu_F(\omega)$ estimates to what extent the fuzzy

set is not empty ; we have $h(\pi) = 1 - m(\emptyset)$, if m is associated to π via (11).
It is always possible to transform a subnormalized basic probability assignment m into a normalized one m′ when $m(\emptyset) \neq 0$. Let M be the set of possibly sub-normalized basic probability assignment, and M* the set of normal ones. A normalization mapping [49] is a mapping $\Phi : M \to M^*$ that reallocates $m(\emptyset)$ to some non-empty subsets of Ω. Such mappings will prove useful further on, to restaure normality after a combination process. The most commonly encountered normalization mapping is the linear one, i.e. $\Phi(m) = \frac{m}{1 - m(\emptyset)}$. When we do not completely rely on a given piece information, i.e. we regard the probabilities $m(A_i)$ allocated to the subsets A_i as too high, we may want to diminish these confidence weights and to increase the weight committed to Ω, i.e. the state of total ignorance. This operation is called *discounting* (see [37]) and produces a new basic probability assignment m′ defined by

$$\begin{aligned} \forall A, A \neq \emptyset, A \neq \Omega, \quad & m'(A) = \lambda \cdot m(A) \\ & m'(\Omega) = \lambda \cdot m(\Omega) + 1 - \lambda \qquad \textbf{(12)} \\ & m'(\emptyset) = 0 \end{aligned}$$

Clearly m′ still satisfies the requirements (2) and (3). The smaller λ, the smaller our confidence in the information represented by m, the more important the discounting. For $\lambda = 1$, $m' = m$. Note that in (12) m is supposed to be normal.
In case where m defines a possibility measure, the transformation (12) can be equivalently written in terms of possibility distributions [47]

$$\forall \omega \in \Omega, \pi'(\omega) = \lambda \cdot \pi(\omega) + 1 - \lambda \qquad \textbf{(13)}$$

since $\pi(\omega) = \Sigma_{A:\omega \in A}\, m(A)$. However, another discounting formula [33] seems more natural in the restricted framework of possibility theory, namely

$$\forall \omega \in \Omega, \pi'(\omega) = \max(\pi(\omega), 1 - \lambda) \qquad \textbf{(14)}$$

The possibility discounting formula (14) can be understood in the following way. λ represents our certainty that the information represented by π is correct. $1 - \lambda$ is the possibility that this information is not correct. $\pi(\omega)$ is the possibility that ω represents the actual world according to the source of information. The modified possibility $\pi'(\omega)$ corresponds to the possibility that either the source is correct regarding the possibility of ω or it is wrong, in agreement with the basic formula (7). In other words, any ω is possible as being the actual world at a degree at least equal to $1 - \lambda$; however the ω's which are regarded as the most possible ones according to source remain the same. Note that the expression (13) has a probabilistic flavor, since (13) can be written $\pi(\omega) + (1 - \lambda) - \pi(\omega) \cdot (1 - \lambda)$, which is of the form $P(\mathbf{a} \vee \mathbf{b}) = P(\mathbf{a}) + P(\mathbf{b}) - P(\mathbf{a}) \cdot P(\mathbf{b})$ (**a** and **b** being stochastically independent).
When the information provided by the source is given under the form of an ordinary subset A of Ω (i.e. $\forall B \neq A$, $m(B) = 0$ and $m(A) = 1$, and $\pi(\omega) = 1$ if $\omega \in A$, $\pi(\omega) = 0$ otherwise), the expressions (13) and (14) coincide. This is the case of a

piece of information, precise (A is a singleton) or imprecise (A has at least two elements) which is regarded as uncertain because the source is not fully reliable. Then (13) and (14) coincide with Shafer simple support belief functions [37] focusing on subset A. On the other hand, (13) and (14) correspond to a combination of π and λ by means of a many-valued implication (Reichenbach's for (13), and Dienes for (14) ; see Rescher [35]), so that $\pi'(\omega)$ can be viewed as evaluating the truth of the following statement : if the source is reliable then ω is restricted by π, in a many-valued logic.

Note that a basic assignment m such that $m(\Omega) > 0$ models an unreliable source in the sense that $m(\Omega)$ is the probability that the source leaves us ignorant, while a basic assignment such that $m(\emptyset) > 0$ models a possibly absurd piece of information. For instance if Ω is a time scale and m models the location of a departure date for a trip, $m(\Omega) > 0$ expresses that the departure date is unknown with some probability, while $m(\emptyset) > 0$ means that there is some probability that the trip will not occur. Moreover discounting accounts for erratic sources without questioning their truthfulness. Mendacious sources (i.e. that consciously lie with probability $1 - \lambda$) are such that when A ($\neq \Omega, \emptyset$) is obtained, it means A with probability λ and $\bar{A}$ with probability $1 - \lambda$. It may induce the following transformation on a normal basic probability assignment m :

$$m'(A) = \lambda m(A) + (1 - \lambda)m(\bar{A}) \quad , \quad \forall A \neq \emptyset, \Omega$$
$$m'(\Omega) = m(\Omega) \quad ; \quad m'(\emptyset) = m(\emptyset)$$

Contrastedly (12) means that when A is obtained, it means A with probability λ and anything (i.e. Ω) with probability $1 - \lambda$.

3. Combining Uncertain Pieces of Information : a General Approach

The problem of parallel combination of uncertain pieces of information can be formulated as follows [14] : Given a set of n uncertainty measures $g_1 \dots g_n$ issued from n sources (e.g. n experts, the results of applying n rules in an expert system...), and defined over a set Ω of alternatives, find an uncertainty measure g which performs a consensus (or a selection) among the n sources, in terms of the g_i's. Note that in rule-based systems Ω often has only two alternatives, say **a**, **¬a** (in that case we are looking for a global estimation of the plausibility/certainty of **a**, taking into account the different sources). More generally Ω may gather more than two mutually exclusive alternatives.

3.1. Some Existing Combination Rules

When the g_i's are belief functions, Dempster rule [5] has been advocated by Shafer [37] as being the most reasonable way of pooling evidence. When $n = 2$, this rule combines 2 basic probability assignments m_1 and m_2 into a third one defined by

$$\forall C \subset \Omega, C \neq \emptyset, m(C) = \frac{\Sigma_{A \cap B = C}\, m_1(A) \,.\, m_2(B)}{\Sigma_{A \cap B \neq \emptyset}\, m_1(A) \,.\, m_2(B)} \quad (15)$$

and $m(\emptyset) = 0$. This rule is associative and commutative.
Shafer [38] has indicated that particular cases of Dempster rule were suggested by scholars such as Hooper, Bernoulli and Lambert in the XVIIth and XVIIIth centuries to combine probability measures, on a 2-alternative set $\Omega = \{\mathbf{a}, \neg\mathbf{a}\}$. Hooper's rule is obtained, with $m_i(\mathbf{a}) = p$, $m_i(\Omega) = 1 - p$ for $i = 1,2$. Lambert's rule correspond to the general case, $m_i(\mathbf{a})$ being the chance that source i is faithful and accurate, $m_i(\neg\mathbf{a})$ the chance that it is mendacious, and $m(\Omega)$ the chance that it is careless.
When the g_i's are probability measures, a more usual pooling operation is a convex combination of the g_i's i.e. there are non-negative numbers $\alpha_1 \ldots \alpha_n$, with $\Sigma\, \alpha_i = 1$ such that

$$\forall A \subset \Omega,\, g(A) = \Sigma_{i=1,n}\, \alpha_i \cdot g_i(A) \quad (16)$$

The α_i's reflect the relative reliability of each source of information.The literature dealing with this approach is not very abundant. See Berenstein et al. [1] for a review. They indicate that under the natural requirement that g is a probability measure such that g(A) only depends upon $\{g_i(A),\ i = 1,n\}$, (16) is the only possible consensus rule. Note that Dempster rule (15) although meaningful when g_1 and g_2 are probability measures, does *not* extend (16) ; moreover (16) assumes more information than (15) (i.e. than Hooper and Lambert's rules for instance) about the sources (i.e. the relative reliability weights α_i).
When the g_i's are possibility measures deriving from possibility distributions $\{\pi_i,\ i = 1,n\}$, then fuzzy set-theoretic operations can be used to pool the evidence. Namely, the following pointwise combination rules have been proposed :

$$\pi_\wedge = \mathop{*}_{i=1,n} \pi_i \quad \text{(fuzzy set intersection)} \quad (17)$$

or

$$\pi_\vee = \mathop{\perp}_{i=1,n} \pi_i \quad \text{(fuzzy set union)} \quad (18)$$

with $x \perp y = 1 - (1 - x) * (1 - y)$. $*$ is generally a 'minimum' operation, but there are other possible choice of operators. See Dubois & Prade [10] for a review of existing approaches to fuzzy set aggregations. Families of parametrized operations for combining fuzzy sets have been investigated. There are no such results in other frameworks, to-date.
Lastly in the MYCIN system, certainty factors $CF_i(\mathbf{a})$ are defined on 2-element sets $\Omega = \{\mathbf{a}, \neg\mathbf{a}\}$ by $N_i(\mathbf{a}) - N_i(\neg\mathbf{a})$, where $N(\mathbf{a})$ and $N(\neg\mathbf{a})$ are degrees of belief and disbelief in $\mathbf{a}$ respectively (cf. section 2.3). They are related to possibility

distributions on Ω, since $CF_i(a) = \pi_i(a) - \pi_i(\neg a)$, and verify $CF_i(\neg a) = - CF_i(a)$. They combine, as proposed by Buchanan and Shortliffe [2] into :

$$CF(a) = CF_1(a) + CF_2(a) - CF_1(a) \cdot CF_2(a) \quad \text{if } CF_1(a) \geq 0 \text{ , } CF_2(a) \geq 0$$

$$= CF_1(a) + CF_2(a) + CF_1(a) \cdot CF_2(a) \quad \text{if } CF_1(a) \leq 0 \text{ , } CF_2(a) \leq 0$$

$$= \frac{CF_1(a) + CF_2(a)}{1 - \min(|CF_1(a)|, |CF_2(a)|)} \quad \text{otherwise.}$$

3.2. A Set-Theoretic View on Combination

In order to clarify the situation it is useful to look at combination from a set-theoretic point of view [14]. Set-theoretic operations such as unions and intersections are indeed the basic pooling rules in set theory. Consider the case of two sources of information giving evidence about the value of some variable u under the form of a set. Namely

Source 1 $u \in A \subseteq \Omega$

Source 2 $u \in B \subseteq \Omega$

This type of information is obviously a particular case of evidential knowledge in the sense of Shafer. The choice of a type of consensus rule is clearly a matter of context. There is no theory liable of prescribing a unique way of pooling these two pieces of information. Each of the three attitudes which arise as basic ones from the study of fuzzy set-theoretic operations i.e. conjunction, disjunction, trade-off can prove relevant in some situation, namely:

conjunctive pooling

If the sources are completely reliable and properly interpreted then a reasonable rule is to conclude $u \in A \cap B$. Note that we should have $A \cap B \neq \emptyset$, otherwise it is self-contradictory to claim that the sources are reliable and that they are correctly interpreted.

disjunctive pooling

If the sources are not completely reliable but we have no information about their reliability then a reasonable attitude may be to conclude $u \in A \cup B$. It amounts to assuming that at least one of the sources tells the truth without specifying which one. The gain in confidence is counterbalanced by a loss in precision.

trade-off

An intermediary attitude consists in considering that what is in $A \cap B$ is a more plausible range for u than $A \cup B$ although one should not reject the latter values. Several kinds of trade-offs between $A \cap B$ and $A \cup B$ may be envisaged. For instance, under the assumption of equal reliability, one can define as a consensus the body of evidence $(\mathcal{F},m)$ such that $\mathcal{F} = \{A , B\}$, $m(A) = m(B) = .5$. If we want to stick to possibility measures, one can consider a possibility distribution such as

$\pi = {}^1/_2(\mu_A + \mu_B)$, where μ_A and μ_B are the characteristic functions of A and B, as a model of consensus.It is easy to check that $\pi(\omega) = Pl(\{\omega\})$ in the sense of the body of evidence $(\mathcal{F}, m)$ just defined. Here, combining evidence comes down to trading off uncertainty (pervading precise results) against imprecision (that may be uninformative).

The above classification allows the identification of the type of consensus corresponding to each combination rule mentioned in 3.1 and the introduction of new rules corresponding to the other type(s) of consensus in each framework. Namely, the next sections try to model each type of combination in the framework of each uncertainty theory.

3.3. Combination in the Belief Function Setting

Dempster rule clearly is a conjunctive pooling method, since it reduces to a set intersection when applied to the pooling of source 1 and source 2. Up to a scaling factor in (15), Dempster rule is formally a random set intersection under a stochastic independence assumption [23]. Indeed (15) can be written using a normalization mapping (section 2.4) :

$$\forall C \neq \emptyset,\ m(C) = \frac{m_\cap(C)}{1 - m_\cap(\emptyset)} \qquad (19)$$

with $\forall C,\ m_\cap(C) = \sum_{\substack{A,B \\ A\cap B = C}} m_1(A) \cdot m_2(B)$. The scaling factor $\frac{1}{1 - m_\cap(\emptyset)}$ enables us to recover a basic probability assignment (i.e. $\sum_A m(A) = 1$) which is normalized (i.e. $m(\emptyset) = 0$), while $m_\cap$ may be subnormalized (such that $\sum_A m_\cap(A) = 1$ but with $m_\cap(\emptyset) \neq 0$ as soon as there exists a focal element of m_1 which has an empty intersection with a focal element of m_2). The amount of conflict between the two sources is

$$k(m_1, m_2) = m_\cap(\emptyset) = \sum_{A \cap B = \emptyset} m_1(A) \cdot m_2(B) \qquad (20)$$

The normalization process in (15) consists in eliminating the conflicting pieces of information between the two sources, consistently with the intersection operation. The normalization is very much questionable in the case of strongly conflicting information. Indeed Dempster rule is very sensitive to the input values in the neighborhood of the total conflict situation and is even discontinuous [14]. Moreover, the assumption of stochastic independence between m_1 and m_2 asserts the possibility of observing simultaneously any A and B such that $m_1(A) > 0$, $m_2(B) > 0$ with probability $m_1(A) \cdot m_2(B)$. The sources being reliable, it entails $A \cap B \neq \emptyset$, and $k(m_1, m_2) = 0$. This suggests that the only safe range of situations where Dempster rule applies is when $\forall$ A, B, $m_1(A) \cdot m_2(B) > 0 \Rightarrow A \cap B \neq \emptyset$; i.e. when no normalization is needed. Letting $\mathcal{F}_1$ and $\mathcal{F}_2$ be the set of focal elements, the condition $A \cap B \neq \emptyset$, $\forall A \in \mathcal{F}_1$, $\forall B \in \mathcal{F}_2$ corresponds to a qualitative notion of independence between $\mathcal{F}_1$ and $\mathcal{F}_2$, i.e. given $u \in A \in \mathcal{F}_1$, u can be in any $B \in \mathcal{F}_2$ and conversely. This qualitative notion is already known in mathematics under the name "set-theoretic independence" [40]. Set-theoretic independence between $\mathcal{F}_1$ and $\mathcal{F}_2$ looks as an extreme requirement for accepting Dempster's rule. When it does not hold, there are other combination schemes that can be considered, to cope with the discontinuity problem.

Yager [48] has suggested to interpret the degree of conflict $k(m_1,m_2)$ as a degree of ignorance about the combined result, by allocating this weight to the referential set Ω. Namely we obtain the normal basic probability assignment m´ defined by the normalization mapping :

$$\begin{aligned} \forall A, A \neq \emptyset, A \neq \Omega, \quad & m'(A) = m_\cap(A) \\ & m'(\Omega) = m_\cap(\Omega) + m_\cap(\emptyset) \\ & m'(\emptyset) = 0 \end{aligned} \tag{21}$$

As it can be easily checked, (21) corresponds to a discounting (in the sense of (12)) of the normalized assignment $m(A) = \dfrac{m_\cap(A)}{1 - m_\cap(\emptyset)}$, $A \neq \emptyset$ with $\lambda = 1 - m_\cap(\emptyset)$.

The fact $m_\cap(\emptyset) \neq 0$ is sometimes interpreted as the possible agreement of the two sources *outside* the set of alternatives Ω [46], [32]. In other words, it would mean that Ω is not necessarily an exhaustive set of alternatives and that both sources agree on the possibility that the reality corresponds to an alternative outside Ω. In (21), the point of view is different, Ω is indeed regarded as the exhaustive set of possible alternatives and then $m_\cap(\emptyset) \neq \emptyset$ expresses a disagreement between the sources which leads to disregard the result of the conjunctive combination as covering the *whole* set of possible alternatives. The greatest the degree of conflict, the less reliable the information, the more important the discounting. This discounting procedure avoids the discontinuity problems of Dempster rule.

Instead of allocating the weight $m_\cap(\emptyset)$ to the whole referential Ω, as in (21), we may also think of assigning this weight to the union of the focal elements of m_1 and m_2, in order to focus on the set of alternatives considered by the sources together, in the spirit of the disjunctive consensus. More generally, we may think of reallocating each quantity $m_1(A) \cdot m_2(B)$ (appearing in (15)) to $A \cup B$ as soon as $A \cap B = \emptyset$, i.e. obtaining from $m_\cap$ a basic probability assignment m´´ by means of the following normalization mapping [14]

$$\forall A \neq \emptyset,\ m''(A) = m_{\cap}(A) + \sum_{\substack{C,D \\ C \cap D = \emptyset \\ C \cup D = A}} m_1(C) \cdot m_2(D) \qquad (22)$$

this corresponds to a "local" discounting where only one of the sources giving C and D is wrong when $C \cap D = \emptyset$. However, a strong discrepancy between sources needs to be seriously examined : sometimes combining will be forbidden [16] (because some sources can indeed be proved wrong), or, as already mentioned, some other kind of combination, different from a conjunctive one even if $A \cap B \neq \emptyset$, will be more appropriate because sources are known not to be reliable, for instance. See section 5.

For belief functions, a systematically disjunctive consensus is defined consistently with Dempster rule by [12], [32] :

$$m_{\cup}(C) = \sum_{A \cup B = C} m_1(A) \cdot m_2(B) \qquad (23)$$

(23) is a union of independent random sets and extends the set-theoretic union to belief functions. See the above references for a study of algebraic properties of belief functions under extended set-theoretic operations. (23) is never proposed by Shafer in his book [37], but it is not less reasonable than Dempster rule, from a set-theoretical point of view. Moreover, combining normal bodies of evidence always lead to a normal body of evidence, i.e. normalization is not necessary.

The consensus rule (16) for probabilities is clearly a trade-off operation. It is easily extended to belief functions as

$$\mathrm{Bel}(A) = \sum_{i=1,n} \alpha_i \cdot \mathrm{Bel}_i(A) \qquad (24)$$

where $\sum_i \alpha_i = 1$,and $\alpha_i \geq 0$, because the convex combination of belief functions is still a belief function. In (24), Bel can be changed into m or Pl, without changing the result. (24) is suggested by Shafer [37], chap. 11. However the knowledge of the α_i's is needed, i.e. we must know about the relative reliability of the sources, contrary to (19-23).

An alternative to (24) is to discount the sources first and then combine the resulting belief functions by means of Dempster rule (or one of its modified version). In that case the discounting factors, say λ_i, account for the *absolute* reliability of the sources. See Shafer [37] for the link between (24) and the Discount + Combine method.

Applying (19) and (23) to probability measures equips probability theory with conjunctive and disjunctive pooling rules. But the following points must be noticed :

- Using a set-theoretic point of view, a probability measure is an ill-located "point" (rather than "set" as is a belief function). Intersecting points is either trivial (when it is the same point) or absurd (when the points are distinct). As a consequence pooling two probability measures in a conjunctive way always results in conflicting evidence, (i.e. $k(m_1,m_2) > 0$) even when the two probability measures are identical ! This situation contrasts with the case of

pooling sets. Note that Dempster rule applied to probability measures always yields a probability measure (due to the normalization factor). But the fact that two probability measures are always conflicting has been used to question the validity of Dempster rule in this case [53].
- Pooling two probability measures in a disjunctive way using (23) no longer yields a probability measure. Indeed, the union of two points is a 2-element set. Hence the resulting body of evidence has focal sets which are not singletons. What we get using (23) is a general belief function. This is why, may be, a disjunctive fusion rule has never been discovered in the probabilistic literature.

3.4. Combination of Fuzzy Sets and Possibility Measures

For possibility distributions, as was said earlier, all kinds of consensus rules exist, in an axiomatic setting, as discussed and surveyed at length elsewhere [10]. Families of disjunctive, conjunctive and trade-off rules exist and can be discriminated by the requirement of structural properties, depending upon the situation. Especially (17) and (18) model conjunctive and disjunctive consensus rules respectively. Trade-offs include the weighted averages of the π_i's ($\pi = \sum \alpha_i \pi_i$, with $\sum \alpha_i = 1$).

The maximum and minimum operations are respectively limit cases of disjunctive and conjunctive attitudes. They can be justified on the basis of requirements such as idempotence ($\pi_1 = \pi_2 \Rightarrow \pi = \pi_1 = \pi_2$) and associativity.

Hybrid consensus rules have been laid bare, for instance a fuzzy set combination $\square$ which is invariant under a De Morgan transformation, namely such that $(A \square B)^c = A^c \square B^c$ where 'c' denotes complementation. In terms of degrees of possibility, $\pi_1(\omega)$ and $\pi_2(\omega)$, it translates into the symmetry property $1 - [\pi_1(\omega) \square \pi_2(\omega)] = (1 - \pi_1(\omega)) \square (1 - \pi_2(\omega))$. Although the arithmetic mean satisfies this property, many other interesting operations, which are not means, also do. Operation $\square$ is called a symmetric sum [41] is always of the form

$$a \square b = \frac{f(a,b)}{f(a,b) + f(1-a, 1-b)} \qquad (25)$$

for some function f such that $f(0,0) = 0$. $f(a,b) = a + b$ corresponds to the arithmetic mean ; $f(a,b) = a \cdot b$ corresponds to an associative operation that displays an hybrid behavior : $a \square b \geq \max(a,b)$ when $a > {}^1/_2$, $b > {}^1/_2$ (disjunctive consensus), $a \square b \in [a,b]$ when $a \geq {}^1/_2 \geq b$ (trade- off) and $a \square b \leq \min(a,b)$ when $a < {}^1/_2$, $b < {}^1/_2$ (conjunctive consensus). Moreover (25) is discontinuous when $a = 0$, $b = 1$ (total conflict) as soon as $f(0,1) = 0$. This is not surprizing since the

denominator in (25) is a kind of normalization factor.
What has been said in the general case of basic probability assignments (section 3.3) about discontinuity due to the normalization and the discounting procedures when performing a conjunctive combination still applies to the particular case of possibility distributions. (17) may provide subnormal results ($\sup \pi_\wedge < 1$) and this rule has a normalized version. Namely the counterpart of (19) is

$$\forall\, \omega \in \Omega, \pi(\omega) = \frac{\pi_\wedge(\omega)}{h(\pi_1,\pi_2)} \qquad (26)$$

where $\pi_\wedge(\omega) = \pi_1(\omega) * \pi_2(\omega)$ and $h(\pi_1,\pi_2)$ is the height of the intersection of π_1 and π_2 defined by

$$h(\pi_1,\pi_2) = \sup_{\omega\in\Omega} \pi_1(\omega) * \pi_2(\omega) = \sup \pi_\wedge \qquad (27)$$

MYCIN's rule of combination corresponds to a fuzzy set-intersection based on the operation product, with this normalization factor, on a binary frame, i.e. a 2-element set $\Omega = \{\mathbf{a}, \neg\mathbf{a}\}$ if we interpret the certainty factors $CF(\mathbf{a})$ as $\pi(\mathbf{a}) - \pi(\neg\mathbf{a})$ [14].
The counterpart of the discounting formula (21) is [14]

$$\forall\, \omega \in \Omega, \pi'(\omega) = [\pi_1(\omega) * \pi_2(\omega)] + (1 - h(\pi_1,\pi_2)) \qquad (28)$$

which corresponds to (13) applied to π with $\lambda = h(\pi_1,\pi_2)$. As already explained in section 2.4, we may prefer the discounting formula (14) in case of possibility distributions, which yields

$$\forall\, \omega \in \Omega, \pi'(\omega) = \max\left(\frac{\pi_1(\omega) * \pi_2(\omega)}{h(\pi_1,\pi_2)}, 1 - h(\pi_1,\pi_2)\right) \qquad (29)$$

$$= \max(\pi(\omega), 1 - h(\pi_1,\pi_2))$$

Lastly, note that Dempster rule can also be applied to possibility measures but does not yield (17) even in a normalized form. Actually the only mathematical discrepancy between possibility theory and belief functions is the use of Dempster rule versus the fuzzy set intersection [39].

3.5. Structural Axiomatic Properties of Combination Operations

Several authors [24], [4] have discussed combination operations in terms of requested algebraic properties only : especially commutativity, associativity, idempotence, symmetry. Indeed, a combination law looks all the better if it possesses such nice algebraic properties. Let us discuss them briefly.
Commutativity is good when sources of information are exchangeable. Associativity is not absolutely required ; a weaker property such as quasi-associativity is often sufficient [49] : a combination operation is quasi-associative if and only if there is an associative operation $*$ and an invertible function Φ such that $f(g_1, g_2, \ldots, g_n) = \Phi(g_1 * g_2 \ldots * g_n)$. Then the main advantage of

associativity, i.e. modularity of the combination when integrating the information from a $(n+1)^{th}$ source, remains. Namely if $G^n = f(g_1, g_2, \ldots g_n)$ then $f(g_1 \ldots g_n, g_{n+1}) = \Phi(\Phi^{-1}(G^n) * g_{n+1})$. Examples of quasi-associative operations are arithmetic mean, normalized fuzzy intersections (equation (26)), and variants (21) and (22) of Dempster rule. The inversion of Φ may require some caution in the calculation of the rule. For instance, when using Dempster rule variant (19), the term $m(\emptyset)$ must be memorized because it is needed to reconstruct $m_\cap$ when a third source must be considered. When using the more refined rule (21) the set $\{m(C) \cdot m(D) \mid C \cap D = \emptyset\}$ must be kept for the same reason.

Idempotence is debatable. The typical idempotent operation is the minimum in fuzzy theory. Choosing between the minimum operation and the product in the case of agreeing sources is dictated by the existing links between the sources. The product operation models a reinforcement effect that may be intuitively satisfying when the sources are unrelated enough : i.e. when an alternative in Ω has a low degree of possibility according to each source, the resulting degree of possibility is still lower. Moreover the combination by means of the product is somewhat in agreement with the idea of impossibility in terms of degree of surprise. Indeed if each source regards an alternative as surprizing for independent reasons, it seems natural to conclude that the alternative should be *very* surprizing since we have different reasons for considering it as such. When the information about independence is not available, the minimum rule appears to be more cautious, due to the idempotence property. The min operation corresponds to a logical view of the combination process : the source which assigns the least possibility degree to a given alternative is considered as the best- informed with respect to this alternative. Moreover the minimum operation can cope with redundant information. Concludingly adopting idempotence is really a matter of context.

Cheng & Kashyap [4] have defended the symmetry property, i.e. operations of the form (25). They implicitly reject conjunctive and disjunctive modes of combination, as a consequence. Hence symmetry cannot be used as a universal property.

The closure property is one that is often used without being explicitly stated. It says that if $g_1, g_2, \ldots, g_n$ belong to some representation framework then the result g of the combination also belong to that framework. For instance any probability theory tenant would assume that pooling two probability measures should produce a probability measure. Similarly, in proposing fuzzy set-theoretic operations, Zadeh [51] took the natural requirement that the intersection or the union of two fuzzy sets is still a fuzzy set. This kind of closure assumption is natural once we want to stay within a given mathematical framework.

Some of the disputes between schools of uncertainty modelling are directly related to the closure property. For instance, Shafer [39] argues against the fuzzy set-theoretic consensus rules because none of them can be obtained when pooling two possibility measures by means of Dempster rule. Indeed the following facts are worth noticing

→ applying Dempster rule (15) to two possibility measures does not yield a possibility measure. The nested property of possibility measures is indeed lost when performing the aggregation, while it is preserved using the fuzzy set-theoretic rules. However, the fuzzy set combination law (17) with $*$ = min can be justified as a *particular* random set intersection under a strong dependence assumption [23], [13].

→ applying the trade-off rule (24) to possibility measures Π_i(and not to possibility distributions π_i!) does not yield a possibility measure. Indeed the set of possibility measures is not clcsed under the convex mixing operation ; in fact the set of belief functions is the convex closure of the union of the set of probability measures and the set of possibility measures [12]. As a consequence, (24) is not equivalent to performing a convex combination of the possibility distributions π_i's.

To proceed forward in the discussion about acceptable combination rules for possibility measures, one must realize that the answer to the debate lies in the closure assumption underlying the combination rules. Within possibility theory, where all evidence is assumed to be consonant, fuzzy set-theoretic combination rules are natural. If possibility measures are to be pooled with other kinds of dissonant evidence, then the combination rule must be general enough to account for the variants of uncertainty, i.e. Dempster rule may for instance apply. Note that the result of pooling two possibility measures by Dempster rule is very close to that obtained by performing the intersection of the underlying fuzzy sets by means of the product [22], so that from a practical point of view the debate can be settled as well. To our opinion Shafer is wrong to dispute fuzzy set intersections on the ground that they do not match Dempster rule. Indeed if we put belief functions into a more general setting, Dempster rule can be disputed on the same grounds ; this is the case if belief functions are imbedded in the wider framework of upper and lower probabilities. Then combination rules for upper and lower probabilities that respect the closure property generally differ from Dempster rule and do not produce a belief function out of two belief functions [14].

Lastly, the closure property can be formulated in a more or less drastic way, according to whether we deal with the set-functions or the data that generate them. For instance the unicity of the trade-off rule (16) for probabilities is due to the following assumption : for any event A, the probability of A is a function of $P_i(A)$, $i = 1,n$ only. Wagner [43] has proved that a similar unicity result holds if this condition is applied to belief functions, and enforces (24) as the only possible combination rule. However rule (16) violates the closure property for possibility measures. The following weighted disjunctive combination rule

$$\forall A, \Pi(A) = \max_{i=1,n} \min(a_i, \Pi_i(A)) \qquad (30)$$

is a counterpart of (16) that respects the closure property for possibility measures. In terms of necessity measures $N(A) = 1 - \Pi(\bar{A})$, this combination rule reads

$$\forall A, N(A) = \min_{i=1,n} \max(1 - a_i, N_i(A))$$

and (30) corresponds to the weighted union [11] of the underlying fuzzy sets. It can be proved that this form of combination, using the maximum of possibility measures is the only one that preserves the mathematical properties of possibility measures [19].

A weaker assumption is that combination is performed by aggregating the underlying distributions (probability weights, possibility weights, or basic probability assignments), and that the result must be a distribution of the same kind. Fuzzy set-theoretic operations and Dempster rule are of that kind.

Concluding, any combination rule is justified not only by the context where combination applies, but also by means of a closure property : the set of possibility distributions is closed under fuzzy set operations, the set of belief functions is closed under Dempster rule. The closure property is a useful technical feature, but it must be stated in such a way as to preserve the possibility for various kinds of combinations.

However combination cannot be discussed only in terms of desirable algebraic properties (as done by Hajek [24], Cheng and Kashyap [4], for instance). First too many algebraic properties lead to sterile impossibility results that cannot solve any practical problem, or to unicity results that are delusive because restricting too much the range of combination attitudes. Second the semantics of the numbers to be combined, their meaning, also helps in choosing the proper combination operations. Namely, combination laws should be in agreement with the axiomatics to which degrees of uncertainty obey.

4. Updating with Several Sources

So far, we have not dealt with dissymmetric combination processes such as the updating of uncertain knowledge in the light of new information [29], [6], [15], [17]. This type of combination process always assumes that some a priori knowledge is available and that it is updated so as to minimize the change of belief while incorporating the new evidence. Bayes rule is of that kind. Dissymmetric and symmetric combination methods correspond to different problems and generally yield different results as briefly examplified in the following in the possibilistic case. Let π be the representation of a piece of information we want to update by taking into account the information that $N(A) = \alpha$ where $A \subseteq \Omega$. A symmetric approach will represent this latter information by $\max(\mu_A, 1 - \alpha)$ and perform a conjunctive combination of the form $\pi * \max(\mu_A, 1 - \alpha)$. The basic idea underlying dissymmetric combination processes is to look for a new representation, here π', which is as close as possible to the previous one in some sense to define and which satisfies the constraint corresponding to the new information, here $N(A) = \alpha$. It may lead here to leave π unchanged on A (at least if $\exists\, \omega \in A, \pi(\omega) = 1$) and to modify π into π' such that $\sup_{\omega \notin A} \pi'(\omega) = 1 - \alpha$ on the complement of A ; what is obtained will highly depend on the measure of information closeness [25] which is used.

The Bayesian approach can be applied to update a prior information (obtained from a source *not* considered in the combination process), taking into account *two* new pieces of information and dealing with the two corresponding sources in a symmetric manner. This combination scheme is examined now and a possibilistic counterpart is studied.

4.1. The Bayesian Approach to Symmetric Combination under a priori Knowledge

In the Bayesian approach [3], it is assumed that some a priori probability assignment on Ω is available, say $p : \Omega \to [0,1]$ with $\sum_{\omega \in \Omega} p(\omega) = 1$ and such that $\forall \omega, p(\omega) > 0$. Ω is a set of hypotheses, and there is another set of symptoms (or observations) related to hypotheses, represented by propositions $\mathbf{b}_1$, $\mathbf{b}_2$,... referring to subsets of a set S. Moreover some conditional probabilities $P(\mathbf{b}_1|\omega)$, $P(\mathbf{b}_2|\omega)$, and $P(\mathbf{b}_1 \wedge \mathbf{b}_2|\omega)$ must be known. They represent the probability that $\mathbf{b}_1$, $\mathbf{b}_2$, and the simultaneous occurrence of $\mathbf{b}_1$ and $\mathbf{b}_2$ are respectively observed when ω is true. To alleviate the data gathering burden, the conditional independence of $\mathbf{b}_1$ and $\mathbf{b}_2$ with respect to all ω's is assumed, i.e.

$$P(\mathbf{b}_1 \wedge \mathbf{b}_2|\omega) = P(\mathbf{b}_1|\omega) \,.\, P(\mathbf{b}_2|\omega)$$

And the updated probability $p(\omega|\mathbf{b}_1 \wedge \mathbf{b}_2)$ based on conjointly observing $\mathbf{b}_1$ and $\mathbf{b}_2$ is given by

$$p(\omega|\mathbf{b}_1 \wedge \mathbf{b}_2) = \frac{P(\mathbf{b}_1|\omega) \,.\, P(\mathbf{b}_2|\omega) \,.\, p(\omega)}{P(\mathbf{b}_1 \wedge \mathbf{b}_2)} \qquad \textbf{(31)}$$

This formula can be transformed so as to express $p(\omega|\mathbf{b}_1 \wedge \mathbf{b}_2)$ in terms of $p(\omega|\mathbf{b}_1)$ and $p(\omega|\mathbf{b}_2)$, as done by Ishizuka et al. [26] :

$$p(\omega|\mathbf{b}_1 \wedge \mathbf{b}_2) = \frac{p(\omega|\mathbf{b}_1) \cdot p(\omega|\mathbf{b}_2)}{p(\omega)} \cdot \frac{P(\mathbf{b}_1) \cdot P(\mathbf{b}_2)}{P(\mathbf{b}_1 \wedge \mathbf{b}_2)} \qquad \textbf{(32)}$$

It can be further simplified by noticing that

$$P(\mathbf{b}_1 \wedge \mathbf{b}_2) = \left(\sum_{\omega'} \frac{p(\omega'|\mathbf{b}_1) \cdot p(\omega'|\mathbf{b}_2)}{p(\omega')}\right) \cdot P(\mathbf{b}_1) \cdot P(\mathbf{b}_2)$$

so that (31) can be expressed as

$$p(\omega|\mathbf{b}_1 \wedge \mathbf{b}_2) = \frac{p(\omega|\mathbf{b}_1) \cdot p(\omega|\mathbf{b}_2) / p(\omega)}{\sum_{\omega' \in \Omega} p(\omega'|\mathbf{b}_1) \cdot p(\omega'|\mathbf{b}_2) / p(\omega')} \quad \textbf{(33)}$$

The apparent similarity between (34) and Dempster rule is striking. Indeed define m_i by $m_i(\{\omega\}) = p(\omega|\mathbf{b}_i)$, $\forall\omega$, and $m_i(A) = 0$ if $|A| \geq 2$. Then let

$$m(\{\omega\}) = \frac{p(\omega|\mathbf{b}_1) \cdot p(\omega|\mathbf{b}_2)}{\sum_{\omega'} p(\omega'|\mathbf{b}_1) \cdot p(\omega'|\mathbf{b}_2)} \quad \textbf{(34)}$$

(33) coincides with (34) as soon as the a priori probability is uniformily distributed i.e. $p(\omega) = 1 / |\Omega|$, $\forall\omega$. It might suggest that Dempster rule is subsumed by Bayes rule. In fact this reasoning is fallacious, first because (34) is only a particular case of Dempster rule ; moreover, Dempster rule and Bayes rule are not derived using the same reference model. For instance the framework of belief functions does not involve conditional probabilities to justify Demspter rule. As Shafer [38] points out, the situations when Dempster rule turns out to be a special instance of Bayes rule, from a formal point of view, are rather frequent and he shows other examples of it. It does not question the interest of Dempster rule, because its justification lies in a non-Bayesian model which possesses its own internal consistency. It is weaker than a full Bayesian model because it assumes less knowledge.

There results shed some light on the problem of combining probability measures by means of Dempster rule. Viewing $p(\omega|\mathbf{b}_1)$ and $p(\omega|\mathbf{b}_2)$ as basic probability assignments, it is clear that there is a missing piece of information in (34) that is present in (33) : the background information. In the Bayesian case this background information is embodied in the a priori probability. In the setting of belief functions, the background information appears as a third belief function. In the Bayesian case, the background knowledge is updated on the basis of the two observations $\mathbf{b}_1$ and $\mathbf{b}_2$, in a disymmetric way, as clear from (33). In the case of belief functions, the background knowledge is considered just as another source and combined with the other pieces of information, in a symmetric way,

$$m(\{\omega\}) = \frac{p(\omega|\mathbf{b}_1) \cdot p(\omega|\mathbf{b}_2) \cdot p(\omega)}{\sum_{\omega'} p(\omega'|\mathbf{b}_1) \cdot p(\omega'|\mathbf{b}_2) \cdot p(\omega')} \quad \textbf{(35)}$$

which clearly differs from (34), but coincides with it when $p(\omega)$ is uniform or when a vacuous belief function is used instead. Note lastly that if $p(\omega|\mathbf{b}_1)$ is interpreted as the probability of ω given $\mathbf{b}_1$ *and* the background information, and the same for $p(\omega|\mathbf{b}_2)$, then in the setting of belief functions the combination (34) is unjustified because the pieces of evidence are related. In (34) and (35), the basic assignments coinciding with probability values, are indeed assumed to be unrelated, while in (33) each $p(\omega|\mathbf{b}_i)$ already encompasses the prior p.

Wu et al. [45] have criticized both Dempster rule and the normalized possibilistic conjunctive rule due to the discontinuity effect. However this effect is clearly due to the normalization factor, which also appears in the Bayesian updating method, as clear from (33) and (34). As a consequence slight changes in $p(\omega|\mathbf{b}_1)$ and $p(\omega|\mathbf{b}_2)$ may lead to very drastic changes in $p(\omega|\mathbf{b}_1 \wedge \mathbf{b}_2)$ when $p(\omega|\mathbf{b}_1)$ and $p(\omega|\mathbf{b}_2)$ are severely conflicting.

(33) can be extended to the case of n pieces of evidence, that are conditionally independent with respect to all ω ; namely terms of the form $\frac{p(\omega|\mathbf{b}_1) \cdot p(\omega|\mathbf{b}_2)}{p(\omega)}$ must be changed into $(\prod_{i=1,n} p(\omega|\mathbf{b}_i)) / p(\omega)^{n-1}$.

Note that in (33) the independence of the pieces of evidence $\mathbf{b}_1$ and $\mathbf{b}_2$ is not assumed i.e. $P(\mathbf{b}_1 \wedge \mathbf{b}_2) \neq P(\mathbf{b}_1) \cdot P(\mathbf{b}_2)$. If we add this independence assumption, (32) simplifies into

$$p(\omega|\mathbf{b}_1 \wedge \mathbf{b}_2) = \frac{p(\omega|\mathbf{b}_1) \cdot p(\omega|\mathbf{b}_2)}{p(\omega)} \qquad \textbf{(36)}$$

However the normalization condition $\sum_\omega p(\omega|\mathbf{b}_1 \wedge \mathbf{b}_2) = 1$ induces constraints on the a priori probability p, given $p(\cdot|\mathbf{b}_1)$ and $p(\cdot|\mathbf{b}_2)$. For instance if Ω is a binary set $\Omega = \{\mathbf{a}, \neg\mathbf{a}\}$, there must hold

$$\frac{p(\mathbf{a}|\mathbf{b}_1) \cdot p(\mathbf{a}|\mathbf{b}_2)}{p(\mathbf{a})} + \frac{(1 - p(\mathbf{a}|\mathbf{b}_1)) \cdot (1 - p(\mathbf{a}|\mathbf{b}_2))}{1 - p(\mathbf{a})} = 1$$

Hence $p(\mathbf{a})$ is completely determined by $p(\mathbf{a}|\mathbf{b}_1)$ and $p(\mathbf{a}|\mathbf{b}_2)$ which sounds strange when only $p(\mathbf{a}|\mathbf{b}_1)$ and $p(\mathbf{a}|\mathbf{b}_2)$ are known.

Besides, we may think of computing the updated probabilities $p(\omega|\mathbf{b}_1 \wedge \mathbf{b}_2)$ in two steps : first updating $p(\omega)$ by $\mathbf{b}_1$ and then by $\mathbf{b}_2$. It gives the following computations

$$p(\omega|\mathbf{b}_1) = \frac{P(\mathbf{b}_1|\omega) \cdot p(\omega)}{P(\mathbf{b}_1)}$$

$$p'(\omega) = p(\omega|\mathbf{b}_1) \text{ (new "a priori" information)}$$

$$p(\omega|\mathbf{b}_1 \wedge \mathbf{b}_2) = \frac{P(\mathbf{b}_2|\omega) \cdot p'(\omega)}{P(\mathbf{b}_2)} = \frac{P(\mathbf{b}_2|\omega)}{P(\mathbf{b}_2)} \cdot \frac{P(\mathbf{b}_1|\omega)}{P(\mathbf{b}_1)} \cdot p(\omega)$$

i.e. (31) with the independence assumption between $\mathbf{b}_1$ and $\mathbf{b}_2$, or equivalently (36) ! It is due to the fact that the series updating process is forgetful, i.e. in the updated probability $p'(\omega)$, the previous evidence $\mathbf{b}_1$ is no longer known.

The questionable assumption with this approach is conditional independence that copes with the lack of data of the form $P(\mathbf{b}_1 \wedge \mathbf{b}_2|\omega)$. Of course it can be justified on the basis of the maximum entropy principle [27], but this principle amounts to represent total ignorance by uniformly distributed probability assignments, an assumption which is debatable. Conditional independence could be weakened into a more general decomposability property, namely $P(\mathbf{b}_1 \wedge \mathbf{b}_2|\mathbf{a}) = P(\mathbf{b}_1|\mathbf{a}) * P(\mathbf{b}_2|\mathbf{a})$ for some operation $*$. This weak definition looks more appropriate in the case of subjective probability measures, without any frequentist interpretation. $x * y$ must clearly range from $x * y = \min(x,y)$ (total dependency) to $x * y = \max(0, x + y - 1)$ (mutual exclusiveness). Rauch [34] has proposed a linear interpolation technique between these two extreme situations, keeping $*$ = product as a particular case. However the properties of the Boolean algebra of propositions induce strong constraints on operation $*$, and we have shown [7] that $*$ must be such that $x * y$ and $x + y - x * y$ are simultaneously associative. A result by Frank [21] proves that continuous candidates for the solution belong to a parametrized family of combination operations, with only one parameter. See Dubois [7] for more details. Another solution is to introduce correlation coefficients explicitly, and this is possible with the Bayesian approach as indicated by Wu et al. [45]. However the pooling of dependent information in the belief function framework is still an open problem (but see [13]).

4.2. Combining Possibility Distributions Under a Priori Knowledge

Note that precise prior probabilities may not exist, and a priori information may be available only in terms of upper probability functions. Thus it may be interesting to apply a Bayesian-like approach in uncertainty models other than probabilities, such as possibility measures. Conditional possibility measures should obey the following axiom

$$\Pi(\mathbf{a} \wedge \mathbf{b}) = \Pi(\mathbf{a}|\mathbf{b}) * \Pi(\mathbf{b}) = \Pi(\mathbf{b}|\mathbf{a}) * \Pi(\mathbf{a}) \qquad \textbf{(37)}$$

where $*$ is the minimum operation or the product ; see Dubois and Prade [18] for justifications. Thus we should have

$$\pi(\omega|\mathbf{b}_1 \wedge \mathbf{b}_2) * \Pi(\mathbf{b}_1 \wedge \mathbf{b}_2) = \Pi(\mathbf{b}_1 \wedge \mathbf{b}_2|\omega) * \pi(\omega) \qquad \textbf{(38)}$$

If the variables underlying the event $\mathbf{b}_1$ and $\mathbf{b}_2$ are non-interactive (i.e. there is no known relation linking these two variables) we have the following decomposability property [9]

$$\Pi(\mathbf{b}_1 \wedge \mathbf{b}_2|\omega) = \min(\Pi(\mathbf{b}_1|\omega), \Pi(\mathbf{b}_2|\omega)) \qquad \textbf{(39)}$$

Thus applying (38) and (37) again we get

$$\pi(\omega|\mathbf{b}_1 \wedge \mathbf{b}_2) * \Pi(\mathbf{b}_1 \wedge \mathbf{b}_2) = \min(\pi(\omega|\mathbf{b}_1) * \Pi(\mathbf{b}_1), \pi(\omega|\mathbf{b}_2) * \Pi(\mathbf{b}_2)) \quad \textbf{(40)}$$

The quantities $\Pi(\mathbf{b}_1 \wedge \mathbf{b}_2)$, $\Pi(\mathbf{b}_1)$, $\Pi(\mathbf{b}_2)$ can be computed from the a priori knowledge about ω represented by the possibility distribution π on Ω,

$$\Pi(\mathbf{b}) = \sup_{\omega \in \Omega} \Pi(\mathbf{b}|\omega) * \pi(\omega) \qquad (41)$$

where $*$ stands for the min operation or the product. The conditional possibility $\Pi(\mathbf{b}|\omega)$ is also supposed to be known for all ω. It represents to what extent $\mathbf{b}$ is a possible manifestation of the presence of ω, viewed as a cause. Strictly speaking we may have $\sup_\omega \Pi(\mathbf{b}|\omega) \leq 1$ only. It is natural to assume that $\sup_\omega \Pi(\mathbf{b}|\omega) = 1$ if there is at least one cause ω that makes the appearance of $\mathbf{b}$ completely possible ; in other words $\mathbf{b}$ is a completely relevant observation for the set Ω. Indeed, if for instance $\sup_\omega \Pi(\mathbf{b}|\omega) = 0$, it would mean that it is impossible to observe $\mathbf{b}$ due to a cause in Ω. In the following we make the assumption $\sup_\omega \Pi(\mathbf{b}|\omega) = 1$ for granted, and call it the postulate of observation relevance, moreover we assume that $*$ is a product. Using (38) and (39), we get

$$\pi(\omega|\mathbf{b}_1 \wedge \mathbf{b}_2) = \frac{\pi(\omega) \cdot \min(\Pi(\mathbf{b}_1|\omega), \Pi(\mathbf{b}_2|\omega))}{\sup_{\omega'} \pi(\omega') \cdot \min(\Pi(\mathbf{b}_1|\omega'), \Pi(\mathbf{b}_2|\omega'))} \qquad (42)$$

$$= \frac{\min(\pi(\omega|\mathbf{b}_1) . \Pi(\mathbf{b}_1), \pi(\omega|\mathbf{b}_2) . \Pi(\mathbf{b}_2))}{\sup_{\omega'} \min(\pi(\omega'|\mathbf{b}_1) . \Pi(\mathbf{b}_1), \pi(\omega'|\mathbf{b}_2) . \Pi(\mathbf{b}_2))} \qquad (43)$$

(42) is the counterpart of the Bayesian formula (31), and (43) the counterpart of (33). However, $\Pi(\mathbf{b}_1)$ and $\Pi(\mathbf{b}_2)$ do not simplify in (43) as they do in (33).

If there is no a priori knowledge, the a priori possibility distribution is vacuous, i.e. $\pi(\omega) = 1, \forall\, \omega \in \Omega$. Due to the postulate of observation relevance, $\Pi(\mathbf{b}_1) = \Pi(\mathbf{b}_2) = 1$, using (41). Note that we may have $\Pi(\mathbf{b}_1 \wedge \mathbf{b}_2) < 1$. It means that $\mathbf{b}_1$ and $\mathbf{b}_2$ are relevant observations for Ω. Then (43) simplifies into:

$$\pi(\omega|\mathbf{b}_1 \wedge \mathbf{b}_2) = \frac{\min(\pi(\omega|\mathbf{b}_1), \pi(\omega|\mathbf{b}_2))}{\sup_{\omega'} \min(\pi(\omega'|\mathbf{b}_1), \pi(\omega'|\mathbf{b}_2))} \qquad (44)$$

Thus, the normalized version of the possibilistic conjunctive combination rule can be viewed as the counterpart of the Bayesian pooling formula (33) in the case of vacuous a priori information while it is also a counterpart of Dempster rule. Assuming a decomposability property based on the product rather than on min as in (35) (which corresponds to a weak form of interactivity) would lead to a formula analogous to (44) with the product instead of min. Formula (44) with the product instead of min is associative, and was first proposed by Smets [42] ; if moreover Ω contains only 2 elements $\mathbf{a}$ and $\neg\mathbf{a}$, it is exactly MYCIN [2]'s

combination formula as said earlier [14]. The behavior of (44) (with $*$ = min or product) is very similar to Dempster rule [14], [39], but it is only quasi-associative (with $*$ = min).
Lastly note that the case of a priori ignorance in possibility theory ($\forall\omega, \pi(\omega) = 1$) leads us to state results very close to Edward's [20] notion of likelihood, especially the identity $\pi(\omega|\mathbf{b}_i) = \Pi(\mathbf{b}_i|\omega)$ and its more general form $\pi(\omega|\mathbf{b}_i) = \dfrac{\Pi(\mathbf{b}_i|\omega)}{\sup_{\omega'} \Pi(\mathbf{b}_i|\omega')}$. Indeed, we have $\Pi(\mathbf{b}_i|\omega) \cdot \pi(\omega) = \pi(\omega|\mathbf{b}_i) \cdot \Pi(\mathbf{b}_i)$, hence $\Pi(\mathbf{b}_i|\omega) = \pi(\omega|\mathbf{b}_i)$, if $\Pi(\mathbf{b}_i) = 1$, for i = 1,2.

5. Concluding Discussion

The combination of various pieces of information supposes that their respective representations can be put in a common mathematical framework. For instance, the combination of probabilistic and possibilistic information can be performed in the framework of Shafer's evidence theory. This paper has proposed a classification of existing combination rules in each framework, and shown that the same kinds of operations exist in the theories of evidence, and possibility. But the solution to a combination problem does not only reside in the choice of a suitable combination rule. Questions such as the reliability of the sources, the possible dependencies between sources and the existence of conflicts between the supplied pieces of information, should be also addressed.
The various levels of reliability of erratic sources can be dealt with either by discounting the information provided by each source or by using a weighted combination law. The first approach supposes that we know the absolute reliability of each source in terms of the probability λ (formulas (12) and (13)) or the certainty (necessity) λ (formula(14)) that the source works. The second method corresponds to a view of the various levels of reliability in a relative way. This is examplified by formula (16) in the probabilistic framework. As explained elsewhere [11], the natural manner for introducing weights λ_i in a min combination operation is $\min_i \max(x_i, 1 - \lambda_i)$ where x_i corresponds to the information provided by source i and λ_i is its relative "ordinal" reliability and we have the scaling condition $\max_i \lambda_i = 1$. For $\lambda_i = 1$ we recover the min operation, for $\lambda_i = 0$ the information is ignored. But the above expression can also be viewed as a min-combination of pieces of information which are discounted in the sense of (14) ; thus the two approaches coincide in that case, up to the normalization of the λ_i's.
Many combination laws which have been reviewed, implicitly or explicitly assume

the independence of the sources in a way or another. Indeed Dempster's rule is basically a random set intersection under a stochastic independence assumption ; the Bayesian approach (see section 4.1) makes use of a conditional independence hypothesis. The latter approach applied to possibility measures enables us to derive the min combination rule provided a non-interactivity requirement (see (39)). Non-interactivity means absence of known dependency, rather than assumption of independence. Indeed in some applications, we ignore whether the pieces of information provided by two sources are or are not derived from completely independent observations. This is why the min operation copes with redundant information due to its idempotence.

We may sometimes have knowledge about the way the pieces of information were derived. This is the case in rule- based systems. The reader is referred to Dubois and Prade [16] for an extensive discussion of how to handle dependencies between variables in the context of rule-based inference systems using a possibilistic approach ; it may also happen that two sources are dependent in another sense, if their behaviors are linked, i.e. for instance one of the sources provides a precise (resp. imprecise) information as soon as the other does the same [13]. Another case of dependency is when one source gives information which is *specific* of the case under consideration while what the other source furnishes is derived from a *general* law (which is somewhat uncertain in the sense that it may have some exceptions). The two pieces of evidence then do not correspond to the same *reference class* [28]. In that case the difference of origin between the two pieces of information can be taken into account in the following way using a possibilistic representation. Let π_1 correspond to the specific information and $\pi_2 = \max(\pi'_2, 1 - \lambda)$ be the representation of the other information where the discounting factor λ reflects the uncertainty attached to the general law. The information supplied by the general law should be questioned when it is not consistent with the specific information. It leads to the following combination [17]

$$\forall\, \omega \in \Omega, \pi(\omega) = \min(\pi_1(\omega), \max(\pi_2(\omega), 1 - h(\pi_1, \pi'_2))) \qquad \textbf{(45)}$$

where in this framework $h(\pi_1, \pi'_2)$ is the natural measure of consistency used as a discounting factor ; see the above-mentioned reference for details. (45) is also called a non-monotonic intersection by Yager [50], when $\lambda = 1$ (i.e. $\pi_2 = \pi'_2$).

Another situation where a non-symmetric combination law is natural is when we have to combine a "new" information with an "old" one. This may be viewed as an updating problem (see the discussion at the beginning of section 4). This may also be viewed, in some cases, as the preference for the "new" piece of knowledge (supposed to be better informed, since more recent) while the "old" information is not totally forgotten to the extent as the "new" information is only certain at the degree λ. This can be expressed in the possibilitic setting as

$$\begin{aligned}\forall\, \omega \in \Omega, \pi(\omega) &= \max(\pi'^{new}(\omega), \min(1 - \lambda, \pi^{old}(\omega))) \qquad \textbf{(46)}\\ &= \min(\pi^{new}(\omega), \max(\pi'^{new}(\omega), \pi^{old}(\omega)))\end{aligned}$$

where the new information is represented by $\pi^{new} = \max(\pi'^{new}, 1 - \lambda)$. When the subsets A and B, such that $\pi^{old} = \mu_A$ and $\pi'^{new} = \mu_B$, are not fuzzy formula (46) allocates a possibility equal to 1 for the elements of B, to $1 - \lambda$ for the elements of A (representing the old information) not compatible with B and to 0 for any other element. When $\lambda = 1$, i.e. the new information is certain, the old one is simply forgotten. This attitude is dual in some sense of the one expressed by (45).

Kyburg [28] has discussed at length the problem of inference from uncertain pieces of information of the form $P(\mathbf{a}|\mathbf{b}_i) \in [\alpha_i, \beta_i]$ pertaining to different reference classes $\mathbf{b}_i$ and having different levels of precision (i.e. $[\alpha_i, \beta_i]$ being anything between a singleton and the unit interval). Kyburg's strategy for choosing the good reference class is summarized by Loui [30] with an Artificial Intelligence point of view. Clearly Kyburg's pioneering work is very relevant for the combination problems in such contexts as rule-based reasoning and data base interrogation and would deserve a fuller discussion.

When the combination law produces an unnormalized result, in a given representation framework (i.e. $\sum p_i < 1$ for a probability distribution, $m(\emptyset) \neq 0$ for a basic probability assignment, $h(\pi) < 1$ for a possibility distribution), we say that the pieces of information are conflicting. If we are absolutely certain that the referential Ω is exhaustive and that the sources are fully reliable (in particular it means that if a source assigns a plausibility equal to zero to an alternative, it is certain that this alternative is impossible ; see Dubois and Prade [14]), it is reasonable to renormalize the result, by dividing it by $\sum p_i$, $1 - m(\emptyset)$ or $h(\pi)$ according to the cases (this presupposes that these quantities are non-zero). If we have some doubt about the global reliability of the sources, we may either discount the result or consider another type of combination law (see sections 3.3 and 3.4).

This paper proposes an overview of the available procedures for combining uncertain pieces of information issued from several sources such as rules in an expert system, sensors, or data bases. It should be distinguished from other combination problems such as the aggregation of multiple criteria or the search for group consensus in decision-making, since the numbers to combine have different semantics and the nature of interesting consensuses may change according to the considered situation (even if the same combination operation appears sometimes in the different contexts). The combination methodology presented in this paper is currently under application to a multiple-source data bank interrogation system [36].

References

1. Berenstein, C., Kanal, L.N. and Lavine, P. (1986) Consensus rules. In L.N. Kanal and J.F. Lemmer (eds.) Uncertainty in Artificial Intelligence, Vol. 1,

North-Holland, Amsterdam, pp. 27-32.
2. Buchanan, B.G. and Shortliffe, E.H. (1984) Rule-Based Expert Systems - The MYCIN Experiments of the Stanford Heuristic Programming Projects, Addison-Wesley, Reading, N.J..
3. Charniak, E. (1983) The Bayesian basis of common sense medical diagnosis, Proc. 1983 American Assoc. Artificial Intelligence Conf., pp. 70-73.
4. Cheng, Y. and Kashyap, R.L. (1989) A study of associative evidential reasoning, IEEE Trans. on Pattern Analysis and Machine Intelligence 11, 623-631.
5. Dempster, A.P.(1967) Upper and lower probabilities induced by a multivalued mapping, Ann. Math. Statistics, 325-339.
6. Domotor, Z. (1985) Probability kinematics conditionals and entropy principle, Synthèse 63, 75-114.
7. Dubois, D. (1986) Generalized probabilistic independence, and its implications for utility, Operations Res. Letters 5, 255-260.
8. Dubois, D. and Prade, H. (1982) On several representations of an uncertain body of evidence. In M.M. Gupta and E. Sanchez (eds.) Fuzzy Information and Decision Processes, North- Holland, Amsterdam, pp. 167-181.
9. Dubois, D. and Prade, H. (1985) (with the collaboration of Farreny, H., Martin-Clouaire, R. and Testemale, C.) Théorie des Possibilités - Applications à la Représentation des Connaissances en Informatique, Masson, Paris. English version "Possibility Theory" published by Plenum Press, New York, 1988.
10. Dubois, D. and Prade, H. (1985) A review of fuzzy set aggregation connectives, Information Sciences 36, 85-121.
11. Dubois, D. and Prade, H. (1986) Weighted minimum and maximum operations in fuzzy set theory, Information Sciences 39, 205-210.
12. Dubois, D. and Prade, H. (1986) A set-theoretic view of belief functions - Logical operations and approximations by fuzzy sets, Int. Journal of General Systems 12, 193-226.
13. Dubois, D. and Prade, H. (1986) On the unicity of Dempster rule of combination, Int. J. Intelligent Systems 1, 133-142.
14. Dubois, D. and Prade, H. (1988) Representation and combination of uncertainty with belief functions and possibility measures, Computational Intelligence 4(4), 244-264.
15. Dubois, D. and Prade, H. (1988) Modelling uncertainty and inductive inference, Acta Psychologica 68, 53-78.
16. D Dubois, D. and Prade, H. (1988) On the combination of uncertain or imprecise pieces of information in rule-based systems, Int. J. of Approximate Reasoning 2, 65-87.
17. Dubois, D. and Prade, H. (1988) Default reasoning and possibility theory, Artificial Intelligence 35, 243-257.
18. Dubois, D. and Prade, H. (1990) The logical view of conditioning and its

application to possibility and evidence theories, Int. J. of Approximate Reasoning 4(1), 23-46.

19. Dubois, D. and Prade, H. (1990) Aggregation of possibility measures, to appear in J. Kacprzyk and M. Fedrizzi (eds.) Multiperson Decision-Making Under Fuzzy Sets and Possibility Theory, Kluwer Academic Pub., Dordrecht, The Netherlands.
20. Edwards, W.F. (1972) Likelihood, Cambridge University Press, Cambridge, U.K..
21. Frank, M.J. (1979) On the simultaneous associativity of f(x,y) and x + y - f(x,y), Aequationes Math. 19, 194-226.
22. Fua, P. (1987) Using probability density functions in the framework of evidential reasoning In B. Bouchon and R.R. Yager (eds.) Uncertainty in Knowledge-Based Systems, Springer Verlag, pp. 103-110.
23. Goodman, I.R. and Nguyen, H.T. (1985) Uncertainty models for knowledge-based systems, North-Holland, Amsterdam.
24. Hajek, P. (1985) Combining functions for certainty degrees in consulting systems, Int. J. Man- Machine Studies 22, 59-76.
25. Higashi, M. and Klir, G. (1983) On the notion of distance representing information closeness : possibility and probability distributions, Int. J. of General Systems 9, 103-115.
26. Ishizuka, M., Fu, K.S. and Yao, J.T.P. (1982) Inference procedure with uncertainty for problem reduction method, Information Sciences 28, 179-206.
27. Jaynes, E.T. (1979) Where do we stand on maximum entropy, In Levine and Tribus (eds.) The Maximum Entropy Formalism, MIT Press, Cambridge, Mass..
28. Kyburg, H.E. (1974) The Logical Foundations of Statistical Inference, D. Reidel, Dordrecht, The Netherlands.
29. Lindley, D.V. (1984) Reconciliation of probability distributions, Operations Research 32, 866-880.
30. Loui, R.P. (1987) Computing reference classes In J.F. Lemmer and L.N. Kanal (eds.) Uncertainty in Artificial Intelligence, 2, North-Holland, Amsterdam, pp. 273-289.
31. Nguyen, H.T. (1978) On random sets and belief functions, J. Math. Anal. & Appl. 65, 531-542.
32. Oblow, E. (1987) O-theory-a hybrid uncertainty theory, Int. J. General Systems 13(2), 95-106.
33. Prade, H. (1985) A computational approach to approximate and plausible reasoning, with applications to expert systems, IEEE Trans. Pattern Analysis & Machine Intelligence 7, 260-283 (Corrections, 7, 747-748).
34. Rauch, H.E. (1984) Probability concepts for an expert system used for data fusion, The AI Magazine 5(3), 55-60.
35. Rescher, N. (1969) Many-Valued Logic, Mc Graw-Hill, New York.

36. Sandri, S., Besi, A., Dubois, D., Mancini, G., Prade, H. and Testemale, C. (1989) Data fusion problems in an intelligent data bank interface, Proc. of the 6th EuReDatA Conference on Reliability, Data Collection and Use in Risk and Availability Assessment, Siena, Italy, (V. Colombari, ed.), Springer Verlag, Belin, 655-670.
37. Shafer, G. (1976) A Mathematical Theory of Evidence, Princeton University Press, N.J..
38. Shafer, G. (1986) The combination of evidence, Int. J. Intelligent Systems 1, 155-180.
39. Shafer, G. (1987) Belief functions and possibility measures. In J.C. Bezdek (ed.) The Analysis of Fuzzy Information, Vol. 1, CRC Press, Boca Raton, Fl., pp. 51-84.
40. Sikorski, R. (1964) Boolean Algebras, Springer Verlag, Berlin.
41. Silvert, W. (1979) Symmetric summation : a class of operations on fuzzy sets, IEEE Trans. on Systems, Man and Cybernetics 9(10), 657-659.
42. Smets, P. (1982) Possibilistic inference from statistical data, Proc. of the 2nd World Conf. on Math. at the Service of Man, Las Palmas, Spain, June 28-July 3, pp. 611-613.
43. Wagner, C.G. (1989) Consensus for belief functions and related uncertainty measures, Theory and Decision 26, 295-304.
44. Walley, P. and Fine, T. (1982) Towards a frequentist theory of upper and lower probability, The Annals of Statistics 10, 741-761.
45. Wu, J.S., Apostolakis, G.E. and Okrent, D. (1990) Uncertainty in system analysis : probabilistic versus, non probabilistic theories, Reliability Eng. and Syst. Safety 30, 163-181.
46. Yager, R.R. (1983) Hedging in the combination of evidence, Int. J. of Information and Optimization Science 4(1), 73-81.
47. Yager, R.R. (1984) Approximate reasoning as a basis for rule-based expert systems, IEEE Trans. Systems, Man & Cybernetics 14, 636-643.
48. Yager, R.R. (1985) On the relationships of methods of aggregation evidence in expert systems, Cybernetics & Systems 16, 1-21.
49. Yager, R.R. (1987) Quasi associative operations in the combination of evidence, Kybernetes 16, 37-41.
50. Yager, R.R. (1988) Prioritized, non-pointwise, non-monotonic intersection and union for commonsense reasoning, In B. Bouchon, L. Saitta and R.R. Yager (eds.) Uncertainty and Intelligent Systems, Lectures Notes in Computer Sciences, n° 313, Springer Verlag, Berlin, pp. 359-365.
51. Zadeh, L.A. (1965) Fuzzy sets, Information and Control 8, 338-353.
52. Zadeh, L.A. (1978) Fuzzy sets as a basis for a theory of possibility, Fuzzy Sets and Systems 1(1), 3-28.
53. Zadeh, L.A. (1985) A simple view of the Dempster-Shafer theory of evidence and its implicatons for the rule of combination, The AI Magazine 7(2), 85-90.

FAILURE RATE ESTIMATION BASED ON DATA FROM DIFFERENT ENVIRONMENTS AND WITH VARYING QUALITY

S. LYDERSEN and M. RAUSAND
SINTEF, Division of Safety and Reliability
N-7034 TRONDHEIM

Summary

Data to be included in reliability databases are typically collected from different sources. The "true" reliability of a component may vary from source to source, due to factors such as different manufacturers, design, dimensions, materials, and operational and environmental conditions. The quality of the data may vary in completeness and level of detail, due to one or more reasons such as data registration methods, company boundary specifications, subjectiveness and skill of the data collector, and time since the failure events occurred.

The paper discusses reliability estimation based on data with the characteristics mentioned above. Special problems and uncertainties are highlighted. The discussion is exemplified with problems encountered during data collection projects such as OREDA.

Introduction

Component reliability data is a necessary input to practically all types of reliability analyses. The sources for such data are, however, often hampered with inconsistencies and other quality problems. When seeking data for a specific component, the analyst often faces one or more of the following problems: The data sources have varying levels of detail and quality. The failure modes and boundary specifications for the component may be ambigously defined. The relevance of the data may be questionable and may also be hampered with confidentiality restrictions.

These problems are described and discussed in this paper. The ideas and results are presented in terms of a constant failure rate in continuous time. With obvious modifications, they also apply for "failure on demand" probabilities.

J. Flamm and T. Luisi (eds.), Reliability Data Collection and Analysis, 243–255.

Some of the topics discussed in this paper are presented by the authors in a wider context in [1].

Data from Different Sources

The "true" reliability of a component will generally be dependent on a number of external and internal factors.

- Component specific factors (design, make, finish etc)
- Environmental factors (humidity, pressure, temperature etc)
- Operation and maintenance factors.

During the data analysis for the OREDA Handbook [2], the variations between the samples were often very significant. The variations in failure rates for a specific drilling equipment is shown in Figure 1. Estimates, together with 90% confidence intervals are presented in Figure 1 for each of the samples. As seen from this figure, the failure rates show significant variations between the samples.

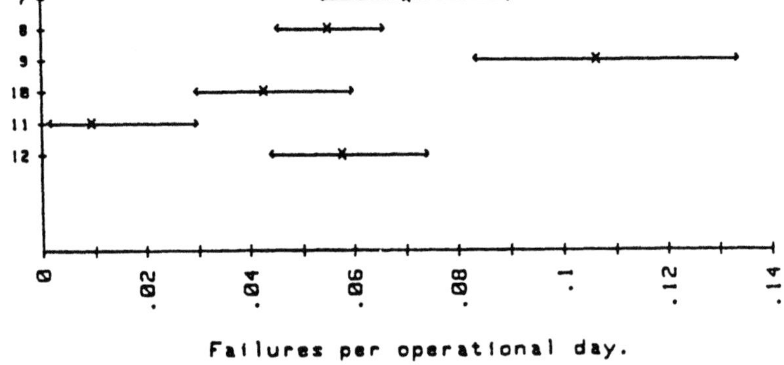

Figure 1. Failure rates and 90% confidence intervals for drilling equipment on 12 different drilling rigs. Plot from the computer program ANEX [3].

In reliability analyses, we usually need a single failure rate estimate for each component and a confidence interval for this estimate. A straightforward approach, although very doubtful, is to pool all the data into one single sample and estimate the failure rate and the confidence interval according to the general formulas for the exponential/Poisson model. This approach will not take the variation between the samples into account. The failure rate estimate will therefore normally be biased and the estimated confidence interval unrealistic narrow.

During the OREDA project, the author M. Rausand suggested a new multi-sample estimator. The estimator has been called the "OREDA estimator". It is based on a Bayesian point of view. The failure rate is regarded as a stochastic variable Λ with one realization per sample. For example, in sample number i, all times to failure are assumed independent, exponentially distributed with failure rate λ_i. The stochastic variable Λ has some probability density $\pi(\lambda)$, with mean

$$\theta = E(\lambda) = \int_0^\infty \lambda \ \pi(\lambda) \ d\lambda, \tag{1}$$

and variance

$$\sigma_\lambda^2 = \mathrm{Var}(\lambda) = \int_0^\infty (\lambda-\theta)^2 \ \pi(\lambda) \ d\lambda. \tag{2}$$

Figure 2 shows a possible prior distribution probability density for the situation in Figure 1.

No specific distribution class, such as the gamma distribution, is assumed for the prior density $\pi(\lambda)$ in the OREDA estimator.

Let X_i denote the number of failures, and τ_i the total time in service in sample number i; $i=1,2,\ldots,k$.

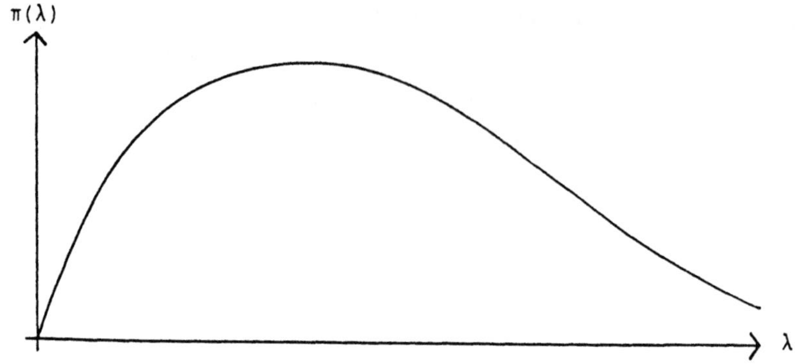

Figure 2. Prior probability density for the failure rate Λ in a sample.

An initial estimate for θ is found by pooling the data:

$$\hat{\theta} = \frac{\sum_{i=1}^{k} X_i}{\sum_{i=1}^{k} t_i} \tag{3}$$

An estimate of the variation between the samples is given by

$$\hat{\sigma}_\lambda^2 = \frac{V - (k-1)\hat{\theta}}{s_1^2 - s_2} \cdot s_1 \tag{4}$$

when this is greater than 0, else 0, where

$$s_1 = \sum_{i=1}^{k} t_i \tag{5}$$

$$s_2 = \sum_{i=1}^{k} t_i^2 \tag{6}$$

$$V = \sum_{i=1}^{k} \frac{(X_i - \hat{\theta} t_i)^2}{t_i} = \sum_{i=1}^{k} \frac{X_i^2}{t_i} - \hat{\theta}^2 s_1 \tag{7}$$

An estimate for the mean failure rate θ is calculated by:

$$\theta^* = \frac{1}{\sum_{i=1}^{k} \frac{1}{\frac{\hat{\theta}}{t_i} + \hat{\sigma}_\lambda^2}} \cdot \sum_{i=1}^{k} \frac{1}{\frac{\hat{\theta}}{t_i} + \hat{\sigma}_\lambda^2} \cdot \frac{X_i}{t_i} \tag{8}$$

An approximate 90% confidence interval for θ is calculated by the formula:

$$\theta^* + \frac{u_{0.05}^2}{2s_1} \pm u_{0.05} \left[\frac{\hat{\theta}}{s_1} + \frac{s_2}{s_1^2} \hat{\sigma}_\lambda^2 + \frac{u_{0.05}^2}{4\, s_1^2} \right]^{1/2} \tag{9}$$

and u_α is the upper 100α% percentile of the standard normal distribution, $u_{0.05} = 1.645$.

Note that this gives a confidence interval for the mean of the prior distribution. It is not a confidence interval for a predicted value of the failure rate. An approximate confidence interval for the failure rate of a component from this item class under similar conditions, is given by

$$\theta^* \pm u_{0.05} \sqrt{\hat{\sigma}_{\theta^*}^2 + \hat{\sigma}_\lambda^2}\ . \tag{10}$$

The OREDA estimator has been thoroughly studied by Spjøtvoll [4], who concluded that the estimator seems to be better than most alternatives. The estimator was used in the OREDA project and is also implemented in the commercially available PC program ANEX [3].

It should be noted that the data illustrated in Figure 1 all originate from drilling rigs in the North Sea with rather similar operational and environmental conditions. The failure rates are significantly different, but the confidence intervals at least cover values of the same size of order. For a number of items in OREDA [2], estimated failure rates showed much more significant discreapancies between the samples.

Unfortunately, few reliability data handbooks or databases include characteristica on the underlying distribution $\pi(\lambda)$. One exception, besides OREDA, is T-boken [5], which contains estimated gamma distribution parameters for $\pi(\lambda)$.

Varying Level of Detail and Data Quality

Practical data collection shows that the level of detail varies between data sources. The main source of reliability information is normally the maintenance recording system and work order forms. The maintenance recording system is primarily designed for follow up and planning of maintenance resources/activities. It is normally not designed to give reliability information. Failure modes are often poorly defined. False alarms and spurious activations in a safety system seldom result in a work order in the maintenance files. Hence, these failure modes are under-represented in several data sources, such as the OREDA Handbook [2].

Boundary specifications often vary from source to source. If we, for example, want to estimate the reliability of a specific type of compressors, problems with the boundaries may arise. For example, the gear box may be part of the compressor's TAG number in some sources. In other sources it may have a separate TAG number, or be included in the TAG number of a more complex unit, such as the motor/turbine. These problems are even more complicated when looking at the control and monitoring system associated to the compressor. A thorough system knowledge, and often some "detective's job", is required to obtain unique boundaries for a specified unit.

Another problem is to obtain detailed inventory data, that is, the number of items of the different kinds/makes, testing schedules, time in operation or standby, modifications etc. This type of information is often poorly recorded and difficult to trace in written sources.

In some sources it may be difficult even to find the number of failures encountered on the item, while other sources may provide detailed failure histories for each labeled component. The normal situation is in between these extremes: The number of failures and a reasonably good failure mode classification is available. Further, the amount of environmental and operational data may vary.

In essence, this means that for some samples, parts of the required data are missing. Typically, some of the environmental/operational data are missing, and some samples will hardly have any failure mode classification. Hence, estimation methods allowing for "missing observations" are needed. This is not a straightforward problem. Typically, in the samples where a specific set of data are missing, these data may have quite different values, compared to samples where these data are available. For a general discussion on this subject, see for example [6].

A data collector may experience everything from neatly updated maintenance files to an operator who states "I don't think we have had more than a few failures on this component for several years". That is, the data may be more or less accurate with respect to operational time, actual number of failures encountered, etc.

The data collector who bases his finding e.g. on maintenance files or interviews, has to use a certain amount of common sense or engineering judgement in the data classification. The subjectiveness and skill of the data collector is a factor which may be difficult to include in the database.

In the IEEE Std 500 Handbook [7], each supplier of data gave a "confidence factor" to account for his evaluation of the data quality.

Established failure event databases are, naturally, never completely error-free. This was also one of the experiences gained during the EuReDatA Benchmark Exercise on Data Analysis performed in 1988 - 1989. It was pointed out that derivation of reliability parameters based on automatic data processing may provide erroneous results due to codification and data inconsistencies, incompleteness of data, and data errors. A sound data analysis should start with a "manual" procedure mainly based on engineering judgement [8].

Varying Data Relevance

Not all the failure history data on an item class may have the same relevance. In many cases, the reliability will have a trend in time. "Old" reliability performance data are less relevant than the more recent.

Varying data relevance also arise for other reasons, which may be illustrated through an example: Consider a user who wishes to estimate the reliability of a given component under specified environmental/operational conditions. Little, if any, data are available on the given item and conditions. To improve the estimation accuracy or feasibility, data on similar items/conditions could be taken into account. If the desired component/conditions is a 4 1/2" XMV (automatic master valve) for a gas production tree, with well head pressure 100 bar, similar items/conditions may be:

- Other well head pressures
- XMV, oil production
- XMV with other dimensions
- Similar gate valves, such as AFV (automatic flow wing valves)

The best way to include data from similar conditions, is by setting up a model for how reliability depends on the relevant factors, such as pressure, GOR (Gas/Oil ratio), size etc. This involves the use of experts in one or more engineering disciplines. In principle, this yields stressor-dependent models. Estimation of parameters in such models is briefly discussed in [1], which also gives further references.

However, in many cases, such modelling is not feasible. The analyst must confine to having little or no data for the desired item/conditions, and some data for similar items/conditions. Some sort of moving average or window estimation technique may prove useful in this situation. This is illustrated in Figure 3 in terms of one factor.

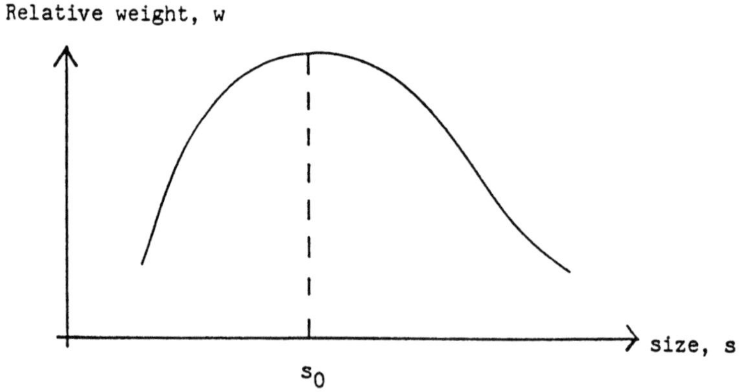

Figure 3. Example on reliability estimation with weights corresponding to data relevance. A reliability estimate is required for the size s_0. Relative weight w is applied for the data during estimation.

The concept of estimation based on debatable evidence provides an alternative way of treating data with varying relevance. It is best explained in terms of an example given by Apostolakis [9]: The problem is to assess the probability of failure to insert the control rods into a nuclear reactor core (to "scram") and, thus, to terminate the neutron chain reaction. A debate went on between the Nuclear Regulatory Commission (NRC) and the Electric Power Research Institute (EPRI) in the United States, as to what the actual field performance data are. This was before the Tsjernobyl incident. The debatable issues were:

- Only one event that could qualify as a scram failure had occured (the Kahl reactor, Germany, 1963). After this incident, scram system modifications have been made, and EPRI claims that it should not be counted.

- Should the number of years of experience include naval reactors and reactors producing plutonium and tritium for use in nuclear weapons, or only commercial and army reactors (as claimed by the NRC)?

- The number of tests per reactor year is a matter of disagreement. The EPRI analysis assumes at least 38 test per year, while NRC is willing to accept only 12 tests per year.

Thus, the statistical evidence was of the form {k events in n tests}, where k is 0 or 1, and n is 7908, 39212, or 114332, depending on one's point of view.

Generally, estimation of the reliability parameter(s) shall be based on statistical evidence E, which is one of the events E_1, E_2, ..., E_r. For example, E_1 = {1 event in 7908 tests}, ..., E_r = {0 events in 114332 tests}. The idea presented by Apostolakis [9] is that the analyst ought to assign probabilities to E_1, ..., E_r, such that $P(E_i)$ represents his belief that E_i is the "true" evidence. It may be argued against this approach that it is, to some extent, based on the subjective belief of the analyst. However, a "conventional" approach, where the analyst states that E_j is the only evidence in which believes, and bases estimation on E_j alone, may be said to be even more subjective. Apostolakis [9] presents a Bayesian estimation method for this situation.

Let θ be the parameter to be estimated, and let $\pi(\theta)$ be its prior distribution. If little is known about θ a priori, a non-informative prior distribution could be used. Bayesian estimation of θ may now be based on the "posterior distribution"

$$\pi(\theta|\underline{E},\underline{P}) = \sum_{i=1}^{r} \pi(\theta|E_i)\ P(E_i), \tag{11}$$

where

$$\underline{E} = (E_1, \ldots, E_r),$$

$$\underline{P} = (P(E_1), \ldots, P(E_r)),$$

$$\sum_{i=1}^{r} P(E_i) = 1,$$

and

$$\pi(\theta|E_i)$$

is the posterior distribution given the data E_i.

<u>Varying Confidentiality</u>

A client who has supplied data to the database, is allowed to read his own data in full detail. Further, he may be allowed to see an "anonymized" version of the rest of the data. For example, the FACTS incident database operated by TNO in the Netherlands contains both restricted and completely accessible information. If the information requested by a client is restricted, it is summarized and anonymized before delivery to the client [10].

<u>Stressor Modelling</u>

Most available reliability data handbooks and databases give little data on how reliability depends on operational and environmental conditions. MIL-HDBK-217E [11] is an exception, where component failure rates are tabled as functions of stressors such as temperature, applied voltage, application, etc. The Nonelectronic Parts Reliability Handbook [12] groups the data according to application, such as "ground mobile", "ground fixed", etc.

Information on reliability dependence on stressors is frequently needed in reliability engineering.

In more advanced reliability databases, it should be possible to estimate the reliability as a function of the environmental and operational conditions. During a comprehensive reliability study of Surface Controlled Subsurface Safety Valves (SCSSV's) performed by SINTEF, a detailed and comprehensive failure event database for such valves was established. To some degree, it was possible to estimate reliability as function of some environmental and operational conditions [13], [14] using proportional hazards modelling (Cox models) [15].

However, the lack of environmental/operational data has made such estimation difficult or impossible in many practical cases.

Reliability Data Dossier

In many practical situations where limited data are available, a reliability data dossier provides a good documentation of the available information. This implies that for each component type included, the analyst must

- Extract reliability data from the sources available to the analyst.

- Write down the extracted data in a systematic way.

- Give a recommended failure rate (or other reliability measures) for the foreseen application and operational and environmental conditions. The recommended failure rate is based on the analyst's judgement of all the available sources.

This approach does not necessarily require sophisticated mathematical methods, but rather good knowledge of the relevant components and applications.

<u>References</u>

1. Lydersen, S.; Sandtorv, H.; Rausand, M.: Processing and Application of Reliability Data. SINTEF Report STF75 A87034, 1987.

2. Offshore Reliability Data Handbook (OREDA). OREDA Participants, Veritec/ PennWell Books, Høvik, Norway, 1985.

3. ANEX ("Analysis of Exponential Life Data"). Computer program for IBM AT or PS/2 developed by SINTEF, Division of Safety and Reliability.

4. Spjøtvoll, E.: Estimation of Failure Rates from Reliability Data Bases. Presentation at the SRE Symposium 1985, Trondheim, September 30 - October 2, 1985.

5. T-boken. Tilförlitlighetsdata för komponenter i svenska kraftreaktorer (Reliability Data for Components in Swedish Power Reactors). RKS - Rådet för Kärnkraftsäkerhet (The Swedish Council for Nuclear Power Safety), 1985.

6. Bayarri, M. J.; de Groot, M. H.: A Bayesian View of Weighted Distributions and Selection Models. In Clarotti, C. A. & Lindley, D. V. (eds.) Proceedings from the course Accelerated Life Testing and Experts' Opinions in Reliability, Lerici, Italy, July 28th - August 1st, 1986. North Holland Publishing Company/Elsevier, Amsterdam, 1988.

7. IEEE Std 500-1984. IEEE Guide to the Collection and Presentation of Electrical, Electronic, Sensing Component, and Mechanical Equipment Reliability Data for Nuclear Power Generating Stations. The Institute of Electrical and Electronic Engineers / Wiley, New York, 1984.

8. Pamme: Contibution to the EuReDatA Benchmark Exercise on Data Analysis. Interatom Report no 70.04440.7, 22 September 1988.

9. Apostolakis, G.: Expert Judgment in Probabilistic Safety Assessment. In Clarotti, C. A. & Lindley, D. V. (eds.) Proceedings from the course "Accelerated Life Testing and Experts' Opinions in Reliability", Lerici, Italy, July 28th - August 1st, 1986. North Holland Publishing Company/ Elsevier, Amsterdam, 1988.

10. Bockholts, P.: Collection and Application of Incident data. In Aniello, A. & Keller, A. Z. (eds.) Reliability Data Bases. Proceedings of the ISPRA Course held at the Joint Research Centre, Ispra, Italy, 21-25 October 1985 in collaboration with EuReDatA. D. Reidel Publishing Company, Dordrecht, Holland, 1987.

11. MIL-HDBK-217E. Military Handbook for Reliability Prediction of Electronic Equipment. Department of Defence, Washington DC, 1986.

12. NPRD-2. Nonelectronic Parts Reliability Data - 2. Rome Air Development Center, New York, 1981.

13. Molnes, E.; Rausand, M.; Lindqvist, B.: Reliability of Surface Controlled Subsurface Safety Valves. Phase II - Main Report. SINTEF Report STF75 A86024, 1986.

14. Tjelmeland, H.: Regresjonsmodeller for sensurerte levetidsdata, med anvendelse på feildata for sikkerhetsventiler i olje/gass produksjons-brønner (Regression Models for Censored Lifetime Data, with Application on Failure Data for Safety Valves in Oil/gas Production Wells). In Norwegian. M.Sc. thesis at the Norwegian Institute of Technology, Trondheim, Norway, 1988.

15. Cox, D. R.: Regression Models and Life-Tables (with discussion). J. R. Stat. Soc. B. 34, 1972, 187 - 220.

OPERATION DATA BANKS AT EDF

PIEPSZOWNIK L., PROCACCIA H.
EDF, Direction des Etudes et Recherches
Département REME
25, allée privée, Carrefour Pleyel
F-93206 Saint-Denis Cedex 1

ABSTRACT

A sum up of the EDF feedback organisation is presented. The three main files are described: the event data bank, the incident data bank and the component event data bank.

PRAMBLE

It is well known that the Electricité de France (EDF) has many nuclear power plants. Having to answer to economical and safety requirements, EDF created an operation feedback system for its own units and also for foreign PWR units, hoping to benefit by previous experience and relevant comparisons.

This feedback has enabled EDF:

- to appreciate a priori and a posteriori safety and availability for French units,
- to justify design modifications and new component or circuit operation procedures, the reliability of which does not correspond with recent purposes,
- to justify units operating to maximum times, during which partial unavailability of safeguards can be accepted,
- to optimise test frequency and material preventive maintenance, to define spare parts stocking,
- to survey component aging.

Three data banks are used for this nuclear power plant operation feedback:

- the incident data bank (FI) concerns the foreign PWR units, since 1974,
- the event data bank (FE) concerns the domestic PWR units, since 1978,
- the Reliability Data System (SRDF) is working since 1978.

These three data banks are representative of EDF operation general feedback organisation; separately, each data bank answers to precise targets given by distinguished users, but they complement each other as the following description will show.

J. Flamm and T. Luisi (eds.), Reliability Data Collection and Analysis, 257–263.

1. THE INCIDENT DATA BANK (FI)

This data bank stores:

- all unit operation results since their first commercial operation (Unit Service Factor, Unit Availability Factor, Unit Capacity Factor),
- all operation incidents, component and circuit incidents.

1.1. Data source

* US power plants

American nuclear power plant data was sourced from the following documents:

- ° Licensed Operating Reactors (NUREG-0020) published by U.S. Nuclear Regulatory Commission (NRC). These documents are called "Grey Books" and give operation results and incidents and also the Licensee Event Reports (LER's) concerning the technical specification not respected,
- ° Nuclear Power Experience (NPE) published by the Stoller Corporation under NRC Authority,
- ° Operations, Maintenance Analysis Report System (OMAR). This data bank belongs to Westinghouse and we use it to compare our data banks to eliminate errors.

Nowadays, 109 PWR's are monitored in FI. Only units having a Design Electrical Rating greater than 400 MWe are taken into account.

Given the diversity and complementary nature of data sources and controls, this data bank can be considered complete and reliable: nowadays, more than 28 000 occurrences are stored.

It is of interest to note that the Electric Power Research Institute (EPRI) has a similar data bank.

* Power plants from Japan, GFR, Belgium, Sweden and Switzerland.

Data sources for these nuclear power plants are International Atomic Energy Agency (I.A.E.A.) annual reports. These documents give significant occurrences which represent only about 10% of all occurrences given for American nuclear plants.
Nowadays, 37 units of these countries are followed in FI.

1.2. The data bank

Three forms are used to supply the data bank:

- the identification form is used for each unit. It is created at the unit's first commercial operation,
- the availability form is filled with monthly unit results, that is Unit Service Factor, Unit Availability Factor and Unit Capacity Factor,
- the incident form is filled anytime there is a unit shutdown for any reason (incident, damage, test, maintenance, etc.) or anytime there is a reported incident without a unit shutdown, with or without partial load.

Components concerned with shutdowns or incidents are aggregated either into plant parts or divided into functional systems, or into components.

We have determined fifteen different plant parts:

1. Reactor (without fuel)
2. Primary pumps
3. Steam generators
4. Reactor coolant system
5. Auxiliary systems and reactor safety systems
6. Reactor instrumentation and control
7. Main turbine and its auxiliary systems
8. Feedwater system
9. Electrical production and internal supply
10. Fuel and handling
11. Refuelling
12. Waste treatment
13. Containment structure
14. Other systems
15. Unspecified

Some components, due to their importance, are considered as plant parts (steam generators for instance).

We use thirty seven groups of components. This assignment is the one which was given by U.S.A.E.C. when our file was created (in 1975).

1.3. File utilisation

Once the data bank is created, a lot of outputs can be selected for using the information.

Periodical reports are written: statistical and comparative balances concerning foreign unit annual results. These studies focus our mind on particular problems essential for unit operation and help us with prospective analysis of operation results.

Particular studies (steam generators, primary pumps, ...) are carried out for obtaining particular information (for instance: to observe some component behaviour versus time).

2. THE EVENT DATA BANK (FE)

The event data bank includes all daily information concerning a unit's operation and, particularly independently of operating incidents, environment related events, human errors, safety related occurrences, information given to external people, ...

On the other hand, this data bank is not concerned with statistical data given elsewhere.

To take into account the various event criteria, it was necessary to create two sub-files:
- the first one containsthe properly so called event sheets,
- the second one holds the "follow-up" reports established for particularly interesting occurrences.

2.1. Data sources

Daily telexes sent by the units are the main sources for this data bank. Periodical reports and significant event reports complete this information.

Event sheets are written, controlled and entered into the data bank within three days. All information allowing event description and the consequences is not always available in such a short time.

At present, the data bank is loaded by centralised staff. In the near future, this job will be performed directly by unit staff.

2.2. The event form

Creation criteria for a data sheet are as follows:

- component incident, damage or test,
- operating incident,
- safety related incident or accident (to be declared to safety authority),
- possible safety related incident or accident (to be declared to safety authority),
- environment related incident or accident (liquid waste, gaseous waste, ...),
- man related event (injuries, accidents, contamination, death),
- external information event.

The event form contains the following items:

- event identification (15 bytes) with concerned unit, the event data and a specific number,
- event type, in order to make it possible to categorise the events according to creation criteria,
- initial document allowing the event form creation,
- equipment affected by the event (this equipment is defined by its system and component description),
- situation of the unit, before and after the occurrence concerning the unit and the reactor,
- consequences in terms of unit and system operation, in terms of safety and in terms of personnel and environment problems,
- event causes and circumstances,
- 400 bytes description summarised or represented by six possible keywords.

2.3. File utilisation

Event data bank outputs can be done in two different manners:

- systematic output of a choice event listing according to given criteria during a given period,
- time sharing operation by using several interrogation jobs. This consultation enables us to know the event reports, or follow-up reports, according to different selection.

Let us give a recent example of this file utilisation.

We observe that an important number of unscheduled shutdowns affects French unit operation.

The event data bank has been utilised in this case to look for and to identify systems and components causing the unscheduled shutdowns. A critical comparison with American nuclear power plants (using FI) allowed us to identify the systems and components we had to concentrate on.

3. THE RELIABILITY DATA SYSTEM (SRDF)

The Reliability Data System collects operating failures in nuclear power plants: it is not an operation data bank.

Since 1973, safety authorities have asked EDF to justify reliability and consequently safeguards for systems safety concerning new PWR's.

This demand caused a working group EDF/CEA (Safety Authority) to be established. This group has to build and to set up at the start of the first French PWR's commercial operation, a reliability data bank including principal mechanical and electromechanical components belonging to nuclear power plant safety related engineering features. This data bank had to feed reliability probabilistic models.

Since 1977, this reliability data bank has been extended to components not concerned with unit safety but also those that cause unavailability.

In 1978, the experience began and concerned 6 units.

During this experience, 800 components were followed by unit pair. Progressively, this number will increase and will concern 1100 components:

. 509 valves,
. 92 pumps,
. 30 tanks,
. 6 turbines,
. 4 diesel generators,
. 102 motors,
. 152 breakers,
. 26 transformers,
. etc.

In 1982, the first SRDF operation analysis was done before extending the system to cover all French nuclear power plants. The statistical sample represents 24 reactors x year experience or 150 000 operation hours, and 4000 failures in which 30% concern pumps and 30% valves.

This first analysis showed the difficulty in describing failures when too many modes and causes, for interpreting them, are given, to a failure sheet writer. This is inappropriate to good data processing.

From the real failures which occurred during unit operation, a logical analysis was set up as an event tree (for sequence) and a fault tree (for modes and causes). This procedure gave the writer only 3 to 6 possibilities for logical failure description.

After this analysis, all the sheets in the data bank were revised. Then 12 new units were entered by SRDF. In 1984, all French PWR's entered the SRDF.

3.1. Data bank input

Three forms are used:

- the identification form describes the monitored component (its historical and technical features, its environment, its component list and scheme, its precise limits),
- the failure form is filled in anytime there is a failure concerning the monitored component. Work orders are used for this job directly by unit staff. The form is filled using the "Failure Logical Analysis Guide". The form is verified and put in the SRDF, using a local display. About 350 failure forms are written during a year concerning a unit pair,
- the operating form is filled every year for every monitored component and contains the operating hours number and the demands number concerning this component. These two numbers are expected to be obtained automatically from unit computers.

3.2. Data bank output

- On-site information treatment

  As soon as a failure form input is made, tests are performed to verify information consistency, existence of necessary data, ...

  It is possible to question the data bank using Boolean equations. For instance, it is possible to get a failure list during a given period concerning a given component belonging to a given circuit. This system gives a precise answer to a precise question asked by unit staff.

- Off-site information treatment

  Data coming from units are gathered in a national computer. These data banks can be questioned anytime using a wide range of selection criteria. Some "on demand" output allows very different types of questions such as:
  . reliability parameter calculation concerning a given component population,
  . research of the best adapted statistical law for a given group lifetime,
  . etc.

Every year, the data is completely updated using the information obtained during the past year.

The data is processed to give the following results:
. operating failure rate,
. "on demand" failure rate,
. repair rate,
. unavailability rate,
. confidence level values (90%).

It is interesting to note that the results given by SRDF are still covered by reduced confidence intervals. SRDF can be used for safety studies. It has been actually used for validating safety studies, for assessing design and reliability of 1300 MWe unit parts and for justifying operation procedures. We expect to use these results to help operation, preventive maintenance, test optimisation, ...

4. CONCLUSION

Efforts carried out by Electricité de France for data collection, data processing and power plant operation feedback analysis correspond to the important and ambitious number of French units operating.

These efforts are not only represented by the daily operation monitoring of every unit through the event data bank, by the particular behaviour of safety related components and the availability of units through the SRDF, but also by the significant, or precursor event research through the older, or different nuclear power plants concerned by the incident data bank.

Probably, these efforts are the most important carried out by a licensee in the world, and are an important contribution to the operation of French nuclear units.

RCM - CLOSING THE LOOP BETWEEN DESIGN AND OPERATION RELIABILITY

HELGE SANDTORV
SINTEF Safety and Reliability
N-7034 Trondheim, Norway

MARVIN RAUSAND
Norwegian Institute of Technology, Division of Machine Design
N-7034 Trondheim, Norway

ABSTRACT

Reliability Centered Maintenance (RCM) is a method for maintenance planning developed within the aircraft industry and later adapted to several other industries and military branches. On behalf of two major oil companies, SINTEF has recently adapted the RCM-method for the offshore industry. The paper discusses the basic merits of the RCM-concept and its suitability for offshore application, based on a case study of a gas compression system.

The availability of reliability data and operating experience is of vital importance for the RCM-method. The RCM-method provides a means to utilize operating experience in a more systematic way. The aspects related use of operating experience is therefore addressed specifically.

1. INTRODUCTION

The experiences and opinions presented in this paper are mainly based on a research program on maintenance technology carried out by SINTEF on behalf of the two oil companies Shell and Statoil. The research program is briefly described by P. van der Vet /9/. One of the main objectives of the program has been to adapt the basic *Reliability Centered Maintenance (RCM)* (/1/, /2/) to a practible tool for use in offshore maintenance planning. In order to verify the tool, a case studiy of an offshore system have been carried out to test the potentials of the method, and adjust the tool based on the experience from this case study. Some of the aspects discussed in this paper will probably also be relevant for most industries where safety and cost optimization in operation is of major concern.

The paper summerizes the main steps of our approach, and lists some of the general experiences from the case study of an export compressor.

2. OFFSHORE MAINTENANCE PARTICULARS

For offshore installations in the Norwegian sector the yearly costs of operation and

J. Flamm and T. Luisi (eds.), Reliability Data Collection and Analysis, 265–281.

maintenance is estimated to approximately NOK 20 billion[1] (1988 NOK) towards the end of the 1990'ies. About NOK 12 billion of these costs are maintenance related when operator's own staff, logistics and catering are included. Deferred production is not included in these costs. In terms of lost revenue these cost may also be of significant magnitude if the oil or gas production is shut down. Even more important are the consequences related to safety. The tragic accident at Piper Alpha is the said underlining of this aspect. It is therefore evident that planning and execution of maintenance of these installations is of decisive importance both for safety and economic reasons.

The maintenance strategy and systems used offshore has rapidly developed during the two decades of oil production in the Norwegian waters, but still it seems that maintenance planning and follow-up is more guided by tradition than by a somewhat more systematic approach. In our opinion, the main contributing factors are:

- There has traditionally been an organizational split between the designers (engineering firm) and the owner operating the installations. The engineering companies are not payed for, nor have the proper competence, to look into maintenance at the design stage.

- The maintenance strategies are usually repair oriented and not reliability oriented; Operating experience is seldom systematically utilized.

- In the past, important systems in the oil/gas process has been built with ample redundancy

Although the basic parameters remain unchanged, some new trends affecting the maintenance planners have recently been brought up:

- Increased number of partly unmanned platforms; notably simple wellhead and booster platforms.

- Significant reduction in manning of the larger production platforms.

- Increased economic incentive due to low oil prices (before the Iraq crisis).

In traditional maintenance planning, both offshore and for landbased industries, the selection of tasks are often based on intuitive reasoning, which typically may include the following:

Experience	"We have always done so, hence it must be right"
Recommendations	"We stick to the recommendations from the manufacturer". Could be a wise way to start, but may not be optimal for our operating and environmental conditions.
Overmaintaining	"We maintain as much as we can just to be on the safe side"
Expertice	"We hire a consultant or use some in-house expert". The common problem is that three years later nobody knows on what grounds the "expert" made his/her decisions.

[1] 20 billion NOK = 3 billion US$

Such procedures are generally less than optimal since there is no organized rationale or structure for selecting preventive maintenance (PM) tasks, and, hence, the way of knowing whether the selected tasks are technically correct or represents a wise allocation of resources.

3. MAIN PHASES IN RCM-ANALYSIS

The main objective of RCM is to maintain the inherent reliability which was designed into the system. By the RCM-method we approach maintenance planning by a logical, well-structured process. The net result is a systematic blend of experience, judgement and reliability techniques and -data to identify optimal preventive maintenace tasks and intervals. The RCM concept has been on the scene for more than 20 years, and applied with considerable success within the aircraft industry, military forces, and more recently within power plant industry (atomic, fossile). Experiences from the use of RCM within these industries (see figure 1) show significant cost reductions in preventive maintenance while maintaining, or even improving, the availability of the systems.

INDUSTRY	APPLICATION	SAVINGS
Civil aircrafts	Propulsion engine DC-10	* 50% reduction in shop spare parts * Significant reduction in labour and material cost
	DC-8 & DC-10 aircrafts	* Reduced no. of items with scheduled overhauls. DC-8: 339 items DC-10: 7 items
Navy Ships		
Canada	32 chiller units on 18 ships	* $ 100.000 per year altogether
USA	38 systems on 4 ships (FF-1052 class)	* Reduced PM manhours for one ship: 43% * RCM proved better statistically than conventional maintenance
Nuclear power generation		
Turkey Point Plant	Cooling water system for main reactor	* PM cost reduction: Manhours: 40%, Spares: 30% * Anticipated reduction in corrective maintenance: 30-40% * Significant reduction in downtime * PM tasks reduced from 17 to 5
Duke Power Station	Main feedwater system	* Increased no. of PM tasks as initial PM program was found inadequate
San Onofre Station	Auxiliary feedwater system	* Net decrease in PM tasks * Deletion of PM tasks in favour of CM tasks * Increased use of surveillance testing to monitor system performance

Figure 1. Documented experience with the RCM method

Before the main RCM analysis is started, one should identify those systems where a RCM-analysis may be of benefit compared with the more traditional maintenance planning. The following criteria for selecting applicable systems are recommended:

- The failure effects must be significant in terms of safety, production loss, or maintenance costs.

- The system complexity must be above average.

- Reliability data or operating experience from the actual system, or similar systems, should be available.

Our RCM approach basically consists of the following four phases:

1. Collection of design-, operational-, and reliability data. Definition of system, system boundaries and major input/output functions.

2. Identification of items (subsystems, units) which are significant for safety, production availability of the plant, or have a high maintenance cost. These items are denoted *Maintenance Significant Items (MSI).*

3. Selection of applicable, and cost-effective, maintenace *tasks* and *intervals*, by using the RCM decision logic. Inclusion of these in the PM-program.

4. Collection and analysis of appropriate data during the in-service phase, and revision of the initial decisions, when required.

This process is illustrated in figure 2.

3.1 Initial data

In the initial phase of the analysis, data are collected and processed for utilization in the further analysis. The initial data may later be adjusted based on updated information and experience. This is of particular relevance for the reliability data, which at the outset of the analysis may be scarce, and mainly based on some generic sources, like the OREDA handbook (/13/). The major steps in this process are:

- Acquisition of technical descriptions of the system in order to define system boundaries, break down the system into smaller entities (subsystem, units), and define main functions (e.g. input/output).

- Definition of operational conditions such as performance requirements, operating profile (continous, intermittant), control philosophy, environmental conditions, acess for maintenance, etc.

- Collection of available operating experience and reliability information from systems with similar design and operating conditions (MTBF, MTTR, failure distribution, typical failure modes, maintenance- and downtime costs).

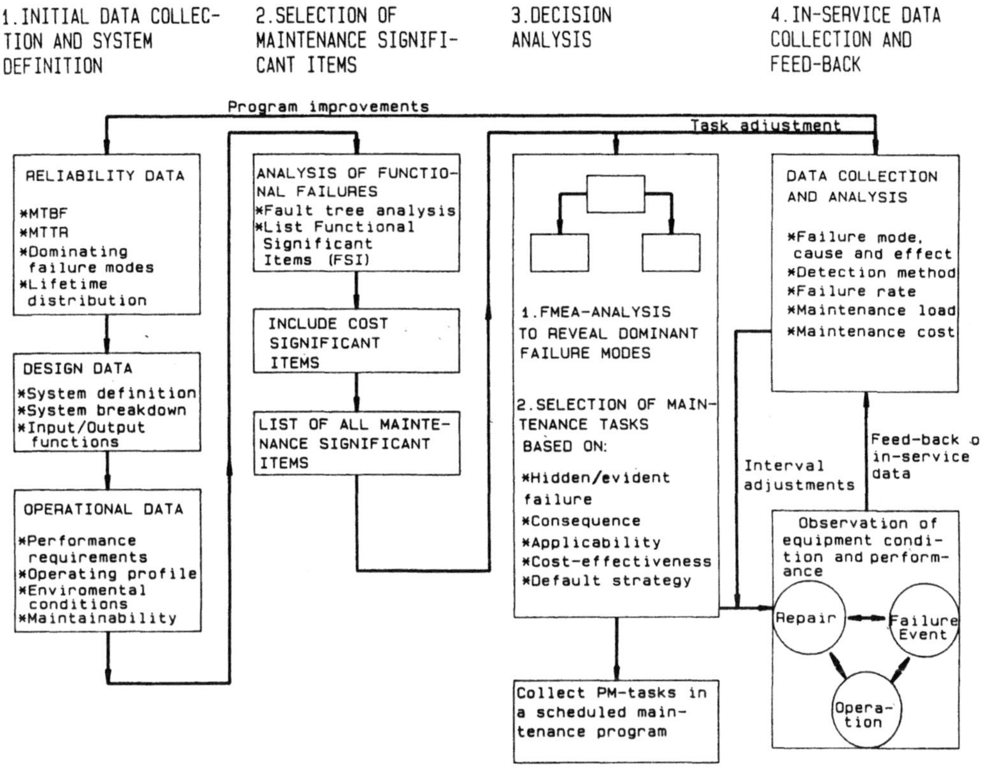

Figure 2. RCM basic modules

3.2 Selection of Maintenance Significant Items (MSI)

There are basically two selection criteria for PM-tasks:

- The effect of loosing one or more of the system functions
- The cost of maintenance in terms of direct cost and downtime cost

An offshore plant consists of a large number of systems, subsystems, and single units that may fail. A main part of these items may fail without consequences which are serious in terms of safety, loss of production, or economic expenditure. These items may be more cost-effective to run until failure, and correct the failure when detected (corrective maintenance). Such items are normally not subjected to a RCM-analysis.

The *Maintenance Significant Items (MSI)* are items with significant failure effects on safety, production availability, or maintenance cost. In order to identify the *MSIs*, we use a *Functional Failure Analysis* normally based on the *Fault Tree Analysis* technique. The Fault Tree Analysis starts with a so-called "TOP" event, which may be a system, or subsystem, failure. The fault tree traces all the causes/failures which may lead to the "TOP" event, by repeating the question "What are the reasons for . . ?" For simple systems, the *MSIs* can be identified directly without a formal analysis.

These items are termed *Functional Significant Items (FSI).*

In addition we identify items with high maintenance cost, low accessibility, long lead time for spare parts, items where external maintenance expertice is required, and add these items to the *FSIs*. The sum then constitute the *MSI*-list (figure 3).

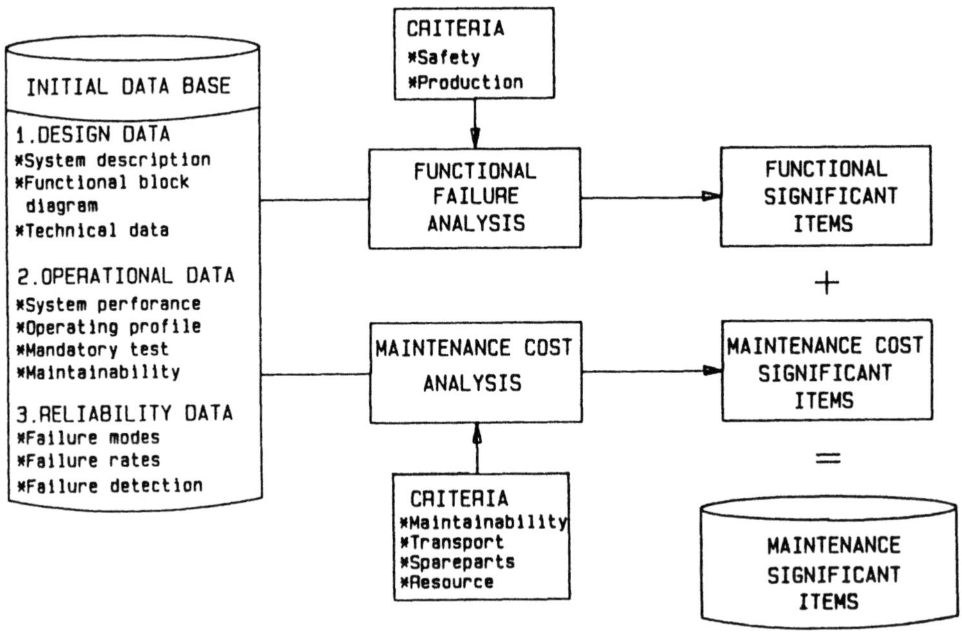

Figure 3. Selection of Maintenance Significant Items

3.3 Selection of maintenance tasks and -intervals

Selection process

This phase is the most novel approach compared to other planning techniques, and use a decision logic to guide the analyst through a question-and-answer process. The input to this analysis is the *MSI*-list defined in the previous phase, together with the data acquired in the first phase. The next step is for each *MSI* to identify those failure modes which are the *dominant* ones, i.e. those failure modes which do not have a very remote probability or with insignificant failure consequences. In addition one should try to identify the potential (or experienced) *cause*, *detectability* (hidden or evident failures), and possible *detection methods.*

The parameters listed above are most systematically identified through a *Failure Mode and Effect Analysis (FMEA)*. In our RCM approach, we use a specific FMEA-form, as shown in figure 4. In our studies we have used specific computer programs for Fault Tree Analysis and FMEA developed by SINTEF Safety and Reliability (/16/). The programs, which run on an IBM AT and PS/2, or compatibles, are very user-friendly and have improved our work efficiency especially on systems with a certain complexity.

RCM - F M E A

SYSTEM : GAS EXPORT COMPRESSOR
SUBSYSTEM :

DATE : 16/01/90
ANALYST : SINTE'/hs
PAGE NO. : 1

ITEM/FAILURE MODE NUMBER	ITEM NAME	MODE OF OPERATION	FUNCTIONAL FAILURE	FAILURE MODE/CAUSE	DETECTION METHOD(S)	FAILURE EFFECTS LOCAL	FAILURE EFFECTS SUBSYSTEM	FAILURE EFFECTS SYSTEM	RELIABILITY DATA	ASSUMPTIONS AND COMMENTS
C1.1	Rotor	Continous	Degraded operation	Vibration (fouling, damages, mis alignment)	Vibration monitoring	Vibration	Undue loads on bearings	Shutdown if vibration excessive	MTBF=10 yrs. MTTR1=50 hrs MTTR2=500hrs	MTTR1=Spare rotor available offshore MTTR2=No spare rotor offshore
C1.2	Rotor	Continous	Low efficiency	Fouling of impeller	Performance monitoring	Increased gas outlet temperature	Increased cooling requirements	Increased power (no. of compressors) required	MTBF=5 yrs. MTTR1=50 hrs MTTR2=500hrs	MTTR1=Spare rotor available offshore MTTR2=No spare rotor offshore
C1.3	Compressor foundation	Continous	Vibrations	Misalignment	Vibration measurements	Wrong shims, loose foundation bolts	Casing vibration, brg. overheating	If excessive shutdown by vibration or brg. temp. alarm	MTBF=10 yrs. MTTR=12 hrs.	MTBF and MTTR estimated
C1.4	Body and internals-exchangeable	Continous	Degraded operation	Material defects, scratches, wear	Internal inspection	Mechanical damage	Variable	From "none" to "shutdown depending on failure	MTBF=10 yrs. MTTR=24 hrs.	MTTR is based on the assumption tha' the failure can be corrected on-site
C1.5	Body and internals-stationary internals	Continous	Degraded operation	Material defects, scratches, wear	Internal inspection	Mechanical damage	Variable	From deferment to shut down depending on failure type	MTBF=20 yrs. MTTR=40 days	MTTR is based on the assumption that the failure cannot be corrected on-site, and no spare available
C1.6	Balance line and piston	Continous	Pressure loss in balance line/piston	External leakage in balance line	Gas detectors and/or inspection. Occasionally thrust brg. temperature	Too low DP across balance piston	Excessive thrust brg. load	Increased thrust brg. tempertures and wear. May result in shutdown.	MTBF=10 yrs. MTTR=12 hrs.	Leakages in balance line system will in addition to increased thrust brg. load also be a safety hazard due to escaping gas
C1.7	Coupling	Continous	Reduced capacity	Defects in material	Inspection	Cracks, deterirated coupling material	Reduced torque capacity	Normally none. Can ultimately result in complete loss of torque	MTBF=4 yrs. MTTR=15 hrs.	

Figure 4. RCM-FMEA form (printout from SINTEF's FMEA program)

Having identified the dominant failure modes and associated parameters, the next step is to perform an analysis based on a decision logic. The scheme we apply is shown in figure 5, and is a guide for the analyst team to verify that the dominant failure modes are identified. The following cases are considered:

- *can be detected* by operating personnel during their normal duties (e.g. watchkeeping, walk-around inspections). (These failures are termed *evident* failures).

- *cannot be detected* as above because the failure do not reveal itself by any physical condition, or because the system is operated intermittantly (e.g. stand-by systems). (These failures are termed *hidden* failures).

- if the failure *develops gradually* and this incipient failure can be detected

- if the failure probability is *age-dependent*, i.e. if there is a predictable wear-out limit

- if the *failure resistance can be reset* to some as-new condition by an appropriate PM-task

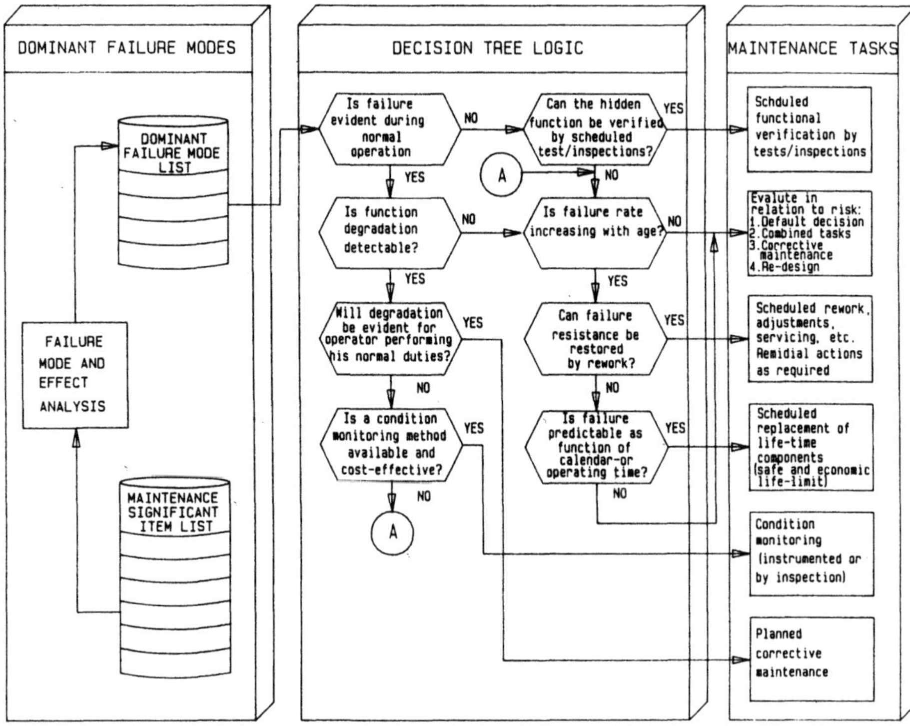

Figure 5. Maintenance Task Assignment

Based on this analysis it should in most cases be possible to arrive at one of the basic maintenance tasks given in the following menu:

1. Scheduled function test

2. Condition monitoring[2]

3. Timebased maintenance, either a scheduled rework task or a replacement task

[2]We use a different definition of CM than used in the aircraft industry. By our definition we mean a task, either manually or by instrumentation, to identify an incipient failure before it develops into a complete functional failure.

4. Planned corrective tasks

   The latter is basically not defined as a task within the RCM-concept, but we have found it useful to include this task as one outcome of the analysis. (See comments later).

   If the amount and/or quality of data acquired during the initial phase of the analysis is not adequate for selecting one of the above four tasks, a fifth category is utilized:

5. Default/evaluation decison

   This "task" means that it is necessary to evaluate this item and failure mode closer, try to acquire additional data, or select a task interval at the outset which is slightly conservative. If the consequence of failure is low, one alternative is "to do nothing", e.g. select corrective maintenance. When a default "task" is selected it is conceived that this strategy should be reviewed as soon as some operating experience is accumulated. These data should then be used to make a new analysis that hopefully will lead to a decision based on more firm knowledge (e.g. a PM-task).

Evaluation of task selection

Two overriding criteria for selecting maintenance tasks are used in RCM:

Applicability: if the task is applicable in relation to our reliability knowledge and in relation to the consequences of failure. If a task is found based on the preceeding analysis, it should satisfy the criteria for applicability

Cost-effectiveness: if a task does not cost more than the failure it is going to prevent

The task selected by the decision logic, and which by definition is the most applicable, should be subject to a final assessment wrt. cost-effectiveness. An applicable task in relation to reliability may not necessarily be the cheapest one, and in this case alternative task/intervals should be re-evaluated. Important aspects to look into here are the possibilities to postpone or advance some tasks in order to group several tasks, co-ordinate smaller tasks, or use any planned (summer) shutdown in order to reduce downtime.

The cost-effectiveness criterion should be emphazised differently depending on the possible failure consequences. For safety important failures, if an applicable task can be found as a result of the decision logic analysis, we have most likely found an acceptable task. For production availability the economic penalities for a complete shutdown is difficult to quantify, as income is not lost, but deferred. If the full loss of revenue is considered, a complete production has to be assessed with a priority close to the safety criteria. For items with mainly maintenance cost as a consequence, the cost-effectiveness criteria will be the dominant one.

3.4 Feed-back of operating data

As mentioned earlier, the reliability data we have access to at the outset of the analysis may be scarce, or even second to none. In our opinion, one of the most significant advantages of RCM is that we systematically analyze and document the basis for our initial decisions, and, hence, can better utilize operating experience to adjust that decision as operating experience data is collected. The full benefit of RCM is therefore only achieved when the "loop is closed" as indicated in figure 2.

Operating experience should be used with basically three objectives which are related to the time span of data collection:

1. Short-term *interval* adjustments

2. Medium term *task* evaluation

3. Long term *revision* of the initial strategy

4. OPERATING DATA

Analysis of operating data

SINTEF Safety and Reliability has acquired a thorough experience on collection and processing of reliability information. We have been an active contractor during all the phases of the OREDA project, and have also established more specific and detailed inhouse databases (/7/). Our experience is that reliability data collection is a very difficult task with a lot of pitfalls. Without detailed knowledge and experience the results from such a task is often of no value.

To optimize a PM-interval for a unit we usually need the following information:

- The time dependent failure rate function wrt. the various failure modes.
- The failure mode distribution for the unit.
- The consequences of the various failure modes, both wrt. safety and economic expenditure.
- How the failure reveal itself, e.g. if the failure develops gradually, and if the failure is evident or hidden

By failure rate function we here mean the intrinsic failure rate function, which is also called the *Force of Mortality (FOM)*. This concept should not be confused with the possibly time dependent failure frequency, which is often called the *Rate of Occurrence of Failures (ROCOF)*. The difference between the *FOM* concept and the *ROCOF* concept is thoroughly discussed by e.g. Ascher & Feingold (/5/). The FOM tells us how fast a certain unit deteriorates, and is thus of significant importance when trying to optimize a PM-interval. The *ROCOF* tells us if there is any trend in the frequency of failures (the *ROCOF*) of a unit which is repaired several times. The *ROCOF* should also be taken into account when trying to optimize a long term PM-plan.

In practical reliability and maintainability studies the two concepts *FOM* and *ROCOF* are often mixed together. The mixing of the two concepts is also clearly seen in many published analyses of reliability data. When times between failures have been recorded, they are very often shifted back to a common starting point, and then analyzed by more or less sophisticated methods like Kaplan-Meier plotting, Hazard Plotting or Total Time on Test (TTT) plotting. These methods are generally very good provided that the input data assumptions are fulfilled. Too often this is not the case.

A repair process can often be modelled as a non-homogeneous Poisson process, and the *ROCOF* may then be estimated as the rate of this process. SINTEF Safety and Reliability has recently developed a computer program for the analysis of non-homogeneous Poisson processes. The program has simply been called **ROCOF** and run on an IBM AT or PS/2 (/18/). The **ROCOF** program utilizes Nelson-Aalen plotting to graphically present the time dependent ROCOF curve. The non-parametrically estimated **ROCOF** curve may be overlaid by a number of parametric curves. The goodness of fit to these curves may be judged by visual inspection. The program also contains two formal statistical tests to test whether the *ROCOF* is constant or not.

If we are lucky and conclude that the *ROCOF* is constant, all the observed times between failures may be shifted back to time zero and analyzed e.g. by the methods mentioned above, Kaplan-Meier, Hazard and TTT plotting. SINTEF Safety and Reliability has also developed a computer program for such analyses. The program which has the same user interface and run on the same type of computers as the ROCOF program is called **SAREPTA** ("Survival and Repair Time Analysis") (/17/).

When the *ROCOF* is not found to be constant, we cannot shift the data back to a time zero and use programs like SAREPTA. If we disregard the non-constant *ROCOF* and run e.g. SAREPTA, we normally arrive at meaningless results. The authors must admit that they have also committed this type of "sin" some years ago before they fully realized the difference between *FOM* and *ROCOF*. We have re-run some of our earlier analyses and have now come to totally different conclusions.

Research is currently carried out to estimate the *FOM* when the *ROCOF* is not constant. This is especially the case when the *ROCOF* is non-constant due to time variations in the environmental and operational conditions, and when the non-homogeneous Poisson process is not a proper model for the repair process.

Experiences with collecting failure data

From our various engagements in the OREDA-project and other data collection projects on offshore installations the common difficulties related to aquiring failure data are:

- Data is generally very *repair* oriented and not directed towards describing failure cause, -mode and -effect

- How the failure was *detected* is raraely stated (e.g. by inspection, monitoring, PM, tests, casual observation). This is a very useful experience to collect in order to select applicable tasks

- *Failure mode* can sometimes be deduced, but this is generally left to the data collector to interpret

- The true *failure cause* is rarely found, but the failure symptom can to some extent be traced

- *Failure effect* on the lower indenture level is reasonably well described, but may often be missing on higher indenture level (system level)

- *Operating conditions* when failure occured is frequently missing or vaguely stated

5. EXPERIENCES WITH THE RCM METHOD

The following summarizes some main benefits, drawbacks and problems encountered during application of the RCM method in some offshore case studies.

General benefits

Cross-discipline utilization of knowledge

To fully utilize the benefits of the RCM concept, one needs contributions from a wider scope of disciplines than what is common practice. This means that an RCM analysis requires contribution from the three following discipline categories working closely together:

1. System/reliability analyst
2. Maintenance/operation specialist
3. Designer/manufacturer

All these categories do not need to take part in the analysis on a full time engagement. They should, however, be deeply involved in the process during pre- and post-analysis review meetings, and quality review of final results. The result of this is that knowledge is extracted and commingled across traditional discipline borders. It may, however, cost more at the outset to engage more personnel.

Tracebility of decisions

Traditionally, PM programs tend to be "cemented". After some time one hardly knows on what basis the initial decisions were made and therefore do not want to change those decisions. In the RCM concept all decisions are taken based on a set of analysis steps, all of which should be documented in the analysis. When operating experience accumulate, one may go back and see on what basis the initial decisions were taken, and adjust the tasks and intervals as required based on the operating experience. This is especially important for initial decisons based on scarce data (e.g. default "tasks").

Recruitment of skilled personnel for maintenance planning and execution

The RCM way of planning and updating maintenance requires more professional skill, and is therefore a greater challenge for skilled engineers. It also provides the engineers with a broader and more attractive way of working with maintenance than what sometimes is common today.

Cost aspects

As indicated, RCM will require more effort both in skill and manhours when first being introduced in a company. It is, however, documented by many companies and organizations that the longterm benefits will far upweigh the initial extra costs. One problem is that the return of investement has to be looked upon in longterm perspective, something that the management not always are willing to take a chance on.

Benefits related to PM-program achievement

Based on the the case studies we have carried out, and experience published by others, the general achievements by RCM in relation to a PM-program can be summerized as follows:

- By careful analysis of the failure consequences, the amount of PM tasks can often be reduced, or replaced by corrective tasks or more dedicated tasks. We have therefore chosen to include corrective maintenance as one task that may be the outcome of the RCM analysis.

- Emphasis has been changed from periodic rework or overhaul tasks of the large assemblies/units to more dedicated object oriented tasks. Consequently, Condition Monitoring were more frequently used to detect spesific failure modes.

- Requirements for spare parts have been reduced as result of better justification for replacements.

- Design solutions were discovered that were not optimal from the safety and plant economic point of view.

Problem areas in the analysis

Lack of reliability data

As indicated the full benefit of the RCM concept can only be achieved when we have access to reliability data for the items being analyzed. Is now RCM worthless if we have no or very poor data at the outset? The answer to this question is no, even in this case the RCM approach will provide some useful information for assessing the type of maintenance tasks. PM intervals will, however, not be available. As a result of the analysis, we should at least have identified the following:

- We know whether the failure involves a safety hazard to personnel, environment or equipment

- We know whether the failure affect production availability

- We know whether the failure is evident or hidden

- We have a better criterion for evaluating cost-effectiveness

The relative importance of reliability data for the RCM analysis is indicated in Figure 6 below:

CRITERIA	RCM with no reliability data	RCM with reliabilitity data
Identification of significant items	Improved. Maintenance planning is focused on functions, not on tasks.	Improved. Maintenance planning focused on functions. Identification of significant items will be more easy and accurate with sufficient data.
Selection of tasks	Improved. Tasks are selected in relation to effect of functional failures and detectability of failure	Significantly improved. Tasks are selected in relation to effect of functional failures, probability of occurance, detectability and failure distribution
Task interval	No improvement	Significant improvement. Task frequencies can be selected optimally.
Analysis of applicability and cost-effectiveness	Slightly improved by the decision logic analysis, i.e. decisions taken ananlytically	Improved. Decisions based on known lifetime distribution and failure detection possibilities.
Traceability of task- and interval decisions	Improved. Rationale of decisions will be documented. This establishes a basis for later optimization.	Significantly improved. Rationale of decisions will be documented and can be optimized based on operating experience.

Figure 6. Relative importance of reliability data for RCM analysis

Criteria for assessing failure consequence

There are three major criteria for the assessment of the consequences of a failure: safety, production availability, and economic loss. In the analysis we have to quantify these measures to some extent to be able to use them as decision criteria. Production availability can be based on a given plant minimum availability number, however, this number may vary depending on delivery contracts and seasonal variations in demand (e.g. gas delivery). Safety is even a more mixed subject. Take one example: We have four firepumps in parallel of which two are sufficient for full firewater supply. If one fails, should this be catgorized as a safety consequence? If we calculate the probability of having a fire and the simultanously probability of not having at least two pumps operational, we will find that the required availability for one pump is very low, and we would not classify this as a safety consequence. Contrary, if a high pressure hydrocarbon gas pipe burst, there is an immediate danger of gas being ignited, and in this case there is an immediate hazard.

The effect could also be a mixed one where the safety criteria has to be weighed against production loss and/or serious economic expenditure. In this case we have used a ranking model where the different criteria are quantified, and the sum of those numbers used to rank the consequence.

Assessing proper interval

The RCM concept is very valuable in assessing the proper *type* of PM task but does not basically include any "tool" for deciding optimal intervals. We have therfore included in the analysis some models and computer codes to assist in this process. The tools we need are basically of two categories:

1. Methods for analysis of in-service operating data discussed in Chapter 4.

2. Methods for cost-optimization of PM intervals based on the result from above and cost of repair in terms of total repair cost and cost of downtime.

Many models for calculation of cost-optimal PM interval exist, but many of them require input data which are not known or only can be assessed with great uncertainty. We have therefore only used very simple models in our calculations, viz. models for assessment of:

- fixed *time* PM, e.g. PM is carried out at fixed intervals even if failure(s) occur between these intervals

- fixed *age* PM, e.g. PM is carried out at a fixed time after a corrective *or* a preventive maintenance task has been carried out

- test interval for equipment with *hidden* failure functions

It is our ambition that once these methods are sufficiently tested and verified as to applicability, they will be integrated as part of the RCM analysis.

6. CONCLUSIONS

RCM is not a single and straight-forward way of optimizing maintenance, but ensures that one does not jump to conclusisons before all the right questions are asked and answers given. RCM can in many respects be compared with Quality Assurance. By rephrasing the definition of QA, RCM can be defined as:

- All *systematic* actions required to plan and verify that the efforts spent on PM maintenance are applicable and cost-effective.

Thus, RCM do not contain any basic new method, but introduce a more structured way of of utilizing the best of several methods and disciplines. Quoting /19/ the author postulates: ...there is more isolation between practitioners of maintenance and the researchers than in any other prefessional activity". We see the RCM concept as a way to reduce this isolation by closing the gap between the traditionally more design related reliability methods, and the more practical related operating- and maintenance personnel.

REFERENCES

/1/ MIL STD 2173 (1986): "Reliability Centered Maintenance. Requirements for Naval Aircraft, Weapon Systems and Support Equipment", Department of Defense, Washington, USA.

/2/ FAA AC 120-17A (1978): "Maintenance Control by Reliability Methods". Advisory Circular, Federal Aviation Administration, Department of Transportation, Washington DC, USA

/3/ OREDA (1984): "Offshore Reliability Data handbook". Published by the OREDA Participants. Available from OREDA, P.O.Box 370, N-1322 Høvik, Norway.

/4/ AIChE (1985): "Guidelines for Hazard Evaluation Procedures", American Institute of Chemical Engineers. New York, USA.

/5/ Ascher,H. & Feingold,H.(1984): "Repairable Systems Reliability, Modeling, Inference, Misconceptions and Their Causes". Marcel Dekker Inc. New York.

/6/ Catola, S.G.(1983): "Reliability Centered Maintenance Handbook". Naval Sea Systems Command S9081-AB-GIB-010/MAINT.

/7/ Molnes,E., Rausand,M. & Lindqvist,B. (1986): "Reliability of Surface Controlled Subsurface Safety Valves". SINTEF report STF75 A86024.

/8/ Moss, M.(1985): "Designing for Minimal Maintenance Expense". Marcel Dekker Inc. New York

/9/ Van der Vet, P.(1989): "Reliability Centered Maintenance". MOU Offshore Conference, Stavanger, Norway (20 Nov. 1989)

/10/ N. Jambulingam, A.K.S. Jardine: "Life cycle costing considerations in RCM. An application to maritime equipment". Reliability Engineering no. 15, 1986.

/11/ B.D. Smith, jr.: "A new approach to overhaul repair work planning". Naval Engineers Journal, 1984

/12/ "Equipment Reliability sets maintenance needs". Electrical World Aug. 1985. (Editorial).

/13/ Brauer: "Reliability Centered Maintenance". IEEE Transaction on reliability, Vol. R.36, No.1, Apr. 1987

/14/ A.M. Smith: " Using reliability centered approach to maintaining nuclear plants". Nuclear Plant Journal, Sept./Oct. 1987

/15/ T.D. Matteson: "Airline experience with reliability cenetered maintenance". Nuclear Engineering and Design 89 (1985).

/16/ CARA (1989): "Computer Aided Reliability Analysis". Computer program for FTA, FMEA and CCA-analysis. Available from SINTEF Safety and Reliability, N-7034 Trondheim, Norway.

/17/ SAREPTA (1989): Computer program for Survival and Repair Time Analysis, for IBM AT & PS/2. Available from SINTEF Safety and Reliability, N-7034 Trondheim, Norway.

/18/ ROCOF (1990): Computer program to analyse repair processes modelled by a non-homogeneous Poisson process, for IBM AT & PS/2. Available from SINTEF Safety and Reliability, N-7034 Trondheim, Norway.^R

/19/ M.A. Malik: "Reliable preventive maintenance scheduleing". AIEE Trans., Vol. 11, pp. 221 -228.

EUREDATA BENCHMARK EXERCISE ON DATA ANALYSIS

A. *BESI*
Commission of the European Communities,
Joint research Centre - Ispra Establishment
Institute for System Engineering and Informatics
21020 Ispra (VA) - Italy

Preface

This report gives a brief overview of the analyses performed during the period 1988-90 within the framework of a EUREDATA Benchmark Exercise (BE) on Data Analysis and of the main results obtained.
Specific reference is made to the problems encountered and the results obtained in the second and conclusive phase of the BE. Furthermore, the major insights on data analysis gained by this BE and lessons learnt are listed and briefly discussed.

1. Introduction: objectives of the BE

Following the programme of EuReDatA to arrive at establishing common guidelines for data collection and analysis, the JRC was requested in April 1987 by the Assembly of the members to organize a Benchmark Exercise on Data Analysis.
The main aim of the BE was the comparison of the methods used by the participants to estimate component reliability from "raw" data (i.e. data collected in the field). The terms of reference of the BE were set up by the JRC coordinators, A.Besi and A.G.Colombo. The reference data set consisted of raw data extracted from the Component Event Data Bank (CEDB), the centralized data bank at the European level which stores data coming from nuclear and conventional power plants located in various European Countries (1).
The first phase of the BE started in June 1988, when a data set in matrix format, stored on floppy disks, was sent to the participants. At the same time the participants received the basic information on the CEDB data structure and coding, necessary for understanding the data.

2. History of the BE; characteristics of the reference data sets

2.1 First phase of the BE (June 1988-September 30th, 1988)

The EuReDatA members which participated in this first phase are: INTERATOM, NUKEM (FRG), VTT (SF), SINTEF (N), ENEA VEL Bologna (I), JRC, ENEA TIB AQ Roma (I),

J. Flamm and T. Luisi (eds.), Reliability Data Collection and Analysis, 283–298.

EDF (F). The participants had agreed to participate on a purely voluntary basis, without any financing by the Commission or EuReDatA.
The reference data set was the CEDB data base related to pumps. It consisted of 450 pumps, which had been monitored for an average period of about 5 years (the observation times were between 3 and 12 years) in 16 European power plant units (10 PWR, 2 BWR, 4 conventional). A total of 1189 failures had been reported on the 450 pumps. According to the CEDB structure, these data included detailed information on the component design and operational characteristics and failure/repair events.
A smaller and more homogeneous data set, a sub-set of the above-mentioned data set, was distributed to the participants at the same time, following the requests of some of them. It consisted of the data related to 20 pumps of the auxiliary feedwater system (named B10 according to the CEDB coding) and 61 pumps of the condensate and feedwater system (F08), which had operated in 12 power plants and had been monitored for periods longer than 3 years. A total of 279 failures had been reported on the 81 pumps.
To guarantee the anonymity of the original data suppliers, some data were partially or totally censored by the JRC staff in preparing the data sets. The IAEA code of the plant was masked, i.e. replaced by an integer. The power value of the plant was cancelled. The utility component and failure identification codes were also masked. In the coded description of the failure the utility codes of "related failures" were cancelled (i.e. the information on linked failures was lost). Moreover, phrases or words with codes used by the utilities were deleted from the free text associated to failures.

An overview of the analyses performed during the first phase of the BE, of the difficulties encountered and the preliminary results obtained, is given in (2). The participants, during their first meeting held in Stockholm on September 30th, 1988, judged the results obtained to be of high interest, though not comparable. This was due to the diversity of the approaches adopted by the participants and the fact that they had occasionally analysed different data subsets, derived from the large reference data set.
The second phase of the BE was launched after the Stockholm meeting. To guarantee that comparable results were obtained by the participants, the terms of reference of the BE were revised as follows (2):

- a smaller reference data set was identified;
- some common minimal objectives for the analyses were indicated (e.g. the estimation of the reliability of the main feedwater pumps).

Even if the main purpose of the BE was comparing the methods of analysis used and not the numerical values obtained, the participants thought that the attainment of comparable results could favour a better understanding of the methods themselves.

2.2 Second phase of the BE and conclusive seminar (January 1989-April 5th, 1990)

The reference data set for the second phase of the BE was distributed in January 1989. It comprised data related to:

- 114 centrifugal pumps, handling water, of the condensate and feedwater system, monitored for a period between 3 and 12 years in 16 European power plants (10 PWR, 2 BWR, 4 conventional);

- 440 failure/repair events reported on the pumps above.

It was a subset of the large reference data set used for the first phase of the BE. It had been suitably revised by the BE organizers to eliminate some inconsistencies of the data which had been detected during the studies of the first phase.
As an aid for data interpretation, the participants received a simplified functional flow-sheet of the condensate and feedwater system of each of the 16 plants.
Most of the pumps are continuously operating, with a few exceptions (see Table 1). In general, in the case of redundant trains, one of the train is, in turn, kept in stand-by, so that the annual operating times of the pumps are fairly balanced.
The set of booster pumps is made of two subsets: the boosters of the extraction pumps and those of the feed pumps.
The boundary of the component "pump", according to the CEDB classification, excludes driver and clutch (1).
Table 2 identifies the subsets of identical pumps and the values of their main design and operating attributes. We note that remarkable differences exist in their engineering and operating attributes, as well as in their operating times (Table 1).
The results of the analyses performed during the second phase of the BE were presented during the second meeting of the participants, held in Siena on March 13th, 1989.
The BE was concluded with a workshop, held at the JRC Ispra on April 5th, 1990 (3).

3. Short description of the approaches adopted by the participants

We refer mainly to the last reports produced by the participants, i.e. to their contributions to the conclusive workshop (3). In these reports most of the participants do not analyse failures on demand, due to the small number of events available in the reference data set.
We do not report on contributions given only by oral presentations.

3.1 INTERATOM

The basic objective of the work, common to all the participants, is the estimation of component reliability from raw data. Nevertheless the main interest of INTERATOM is investigating on the possibility to generate statistical results "of such a quality that they can be used as a basis for decision making processes (e.g. risk assessment for licensing purposes)".
Then, as a first step of its work (4), INTERATOM tries to check quality, consistency and degree of completeness of data. The difficulties of data interpretation are particularly enphasised.
The difficulties of obtaining reliable results from raw data, mainly when the observation period is short and data are heavily censored, are highlighted. The "Mean Time Between Reported Events (MTBRE)" for each feedwater pump is computed. It is shown that the variations of the MTBRE from one pump to the other are surprisingly high (up to a factor of 4) even for pumps pertaining to equal parallel trains in the same plant; i.e., for pumps having the same engineering attributes, submitted to the same operating conditions and monitored with the same criteria of reporting.

TABLE 1

Reference data set for the BE, Phase 2
Number of pumps installed and relative capacity, number of failures occurred, number of individual pumps observed (i.e. replacements included), and cumulative operating times.

	EXTRACTION		FEED		BOOSTER		Number of pumps observed and cumulative operation time (hours)	
Plant	Number of pumps and capacity	Number of failures	Number of pumps and capacity	Number of failures	Number of pumps and capacity	Number of failures		
1 PWR	3 x 50%	13	2 x 50%	14	-	-	5	105400
2 PWR	3 x 50%	13	2 x 50% (+2 repl.)	15	-	-	7	137663
3 PWR	3 x 50%	7	2 x 50%	18	-	-	5	97507
4 PWR	3 x 50%	13	2 x 50%	23	-	-	5	112276
5 PWR	3 x 50%	8	2 x 50%	14	-	-	5	93074
6 BWR	3 x 50% 2 x100% 2 lines	0 4	3 x 50%	18	3 x 50%	3	13	74006
7 CON	2 x100%	11	3 x 50%	28	3 x 50%	13	8	455204
8 CON	2 x100%	7	3 x 50%	27	3 x 50%	10	8	497963
9 CON	2 x100%	7	3 x 50%	12	3 x 50%	10	8	477507
10 CON	2 x100%	10	3 x 50%	14	3 x 50%	11	8	501948
11 BWR	2 x 50%	0	3 x 50%	23	2 x 50%	2	7	275856
12 PWR	3 x 50% 2 lines	6	3 x 50% 2 lines	15	-	-	12	534075
13 PWR	3 x 50%	9	3 x 50%	14	-	-	6	111683
14 PWR	-	-	3 x 50%	1	3 x 50%	8	6	55102
15 PWR	3 x 50%	5	2 x 50%	15	-	-	5	87160
16 PWR	-	-	3 x 50%	7	3 x 50%	12	6	49700
	44	113	47	258	23	69		

Notes:

1. All the pumps have a continuous operating mode,with the following exceptions of pumps kept in stand-by:
   - plant 6, one of the 3 extraction pumps and one of the 3 feed pumps
   - plant 6, one of the 2 pumps of extraction from the condenser of the two main feedwater turbo-pumps
   - plant 13, one of the three feedwater pumps
2. The number of individual pumps observed is obtained by adding to the number of pumps installed the number of the possible replacements. In the plant no 2, two replacements occurred in the two feedwater operating positions.

TABLE 2

Reference data set for the BE, Phase 2.
Identical pump subsets and related design and operating attributes.

Plant No.	Plant Type	Pump Appli-cation	Design Power [kW]	Oper. Flow [m^3/s]	Oper. Press [bar]	Oper. Head [bar]	Oper. Temp. [°C]	Numb. of Pumps	Numb. of Fail
1-5,15	PWR	Extr.	2240	0.490	32.0	31.9	32	18	55
1-5,15	PWR	Feed	3580	0.810	63.0	31.0	180	14	95
6	BWR	Extr.	1365	0.660	15.6	15.5	33	3	0
6	BWR	Boost	2170	0.660	39.0	27.0	34	3	3
6	BWR	Feed	4635	0.800	70.0	36.0	192	3	17
6	BWR	Extr.*	15	0.017	5.9	5.6	33	4	4
7-10	CONV	Extr.	880	0.214	31.4	31.4	35	8	35
7-10	CONV	Boost	100	0.157	15.7	7.8	162	12	44
7-10	CONV	Feed	4217	0.157	220.0	204.0	168	12	80
11	BWR	Boost	294	0.357	9.0	-	29	2	1
11	BWR	Extr.	294	0.357	20.0	-	57	2	-
11	BWR	Feed	442	0.357	76.5	-	135	3	21
12	PWR	Extr.	1200	0.238	30.0	30.0	34	6	5
12	PWR	Feed	3500	0.358	100.0	63.0	115	6	12
13	PWR	Extr.	3125	0.520	47.0	46.0	40	3	9
13	PWR	Feed	4909	0.856	73.5	40.5	183	3	14
14	PWR	Boost	1550	1.000	23.0	13.0	180	3	8
14	PWR	Feed	7270	1.000	77.0	54.0	180	3	1
16	PWR	Boost	445	0.406	18.1	8.5	250	3	9
16	PWR	Feed	3600	0.406	72.0	-	250	3	7

* Extraction from the condenser of the main feedwater turbo pump.

INTERATOM identifies, for statistical inference of reliability parameters, a set of 32 main feedwater pumps of commercial BWR or PWR units, all with similar technical characteristics, observed during a similar period (4 years) from the beginning of their operating life. Then the events related to these pumps are submitted to a thorough analysis, to check the independence between events, the consistency and credibility of the relative coding, etc. At last, from the set of checked events, a failure rate for complete and sudden failures is derived, assuming an exponential lifetime probability distribution. In the opinion of INTERATOM, the hypothesis of constant failure rate is acceptable in system reliability assessment for PSA purposes.
Nevertheless, a demonstration of the actual time dependency of the hazard rate for new pumps is given. INTERATOM considers a set of 18 pumps, with similar engineering and operating attributes, which have been observed from the beginning of their life up to their first external leakage. The failure times are plotted on Weibull probability paper; the approximately linear trend of the plotted points show that these times can be assumed Weibull-distributed. From the graph in Weibull paper the shape parameter ß is graphically estimated to be 1.9, compared with the estimate of 1.8 provided by a least square regression. This estimated value of the shape parameter indicates that the hazard rate is approximately linearly increasing in time.

The usual statistical assumptions made for reliability estimation in the case of repairable components are that successive lifetimes are independent, identically distributed random variables, i.e. the renewal model is "component as good as new after each repair". INTERATOM demonstrates that these assumptions are not justified by the data. For a group of 32 big, continuously operating, pumps the times to the first leakage and the subsequent times to the second leakage are considered. The substantial decrease of the expected time-to-leakage after the first repair is a demonstration of the imperfection of the latter. By the use of TTT plots it is shown that different ageing trends characterize the two periods, the one up to the first failure and the one between the first and the second failure.

3.2 VTT

The main objective of VTT analyses was not to obtain good values for the reliability parameters to be used for specific purposes (e.g. for PSA), but to compare the various methods adopted by the participants for the estimation of these parameters.

As exploratory data analyses (5), they analyse trends in failure frequency. By simply representing along a calendar time scale the events occurred to a component, remarkable differences in the operational behaviour of the components pertaining to the same system and to the same plant appear clearly. Investigation on the coded information on failure causes, failure detection, parts failed, failure descriptors, etc, show that the most significant causes of failure are firstly "normal degradation" (i.e. expected ageing of parts), then "material incompatibility" and "engineering design". Errors in maintenance/testing/setting play also an important role. The most frequent failed parts are the shaft sealings and, at a very lower level of frequency, bearings, shaft and the cooling system.
In (5) VTT defines some simple performance indicators with reference to component availability, reliability and maintainability. These indicators are evaluated for the extraction and feedwater pumps of three identical PWR units; it is computed, for instance, the impact on these indicators of pump piece-part failures. Graphical representations of these indicators, very

easy to understand, highlight the differences in performance existing between plants and individual pumps within the same plant. According to VTT, such a programme, performing very simple descriptive analyses, understandable by persons not having any skill in reliability engineering and statistics, should be regarded as an example of immediate use of collected data to aid the plant operator in monitoring equipment performance, making decisions in maintenance activity, etc.

VTT obtains the subsets of pumps to consider for estimation by combining plant type (PWR, BWR, conventional) with application type (feed, extraction, booster) (6). It is noted that pumps technical characteristics can vary considerably inside each subset.

For the estimation of reliability parameters, two renewal models are considered. Both models assume independence between failures with different failure mode (e.g. sudden failures occur independently of incipient failures) and component "as good as new after repair". The first model, the one adopted by all participants, assumes that the component renewal occurs only during the repairs associated with the failures of the type considered. For instance, if we consider sudden failures, the component is renewed only during the repairs following sudden failures.
The second model also considered by VTT, assumes that the renewal occurs during all the repairs, independently of the type of failure to which each repair is associated. In this case the renewals are more frequent along the component operating history. As a result, if we consider failures of a specific type, e.g. sudden, we have a remarkable number of additional censored lifetimes: i.e. all the lifetimes ending with repairs associated with incipient failures. Also, the times to failure are consequently shortened.

For the estimation of the expected failure, repair and restoration times various distributions are consided. These are exponential, Weibull, log-normal, mixture of two exponential, conditional exponential and gamma. As failure time distributions are chosen those which maximize the likelihood function, while the repair and restoration time distributions are chosen on the basis of the Kolmogorov-Smirnov test of the goodness of fit. In fact it is aknowledged that the use of the latter test may be misleading when the data contain censored observations.
As to the effect of renewal model assumed for sudden failures, the mean time to failure turns out to be longer in the case of lifetime censored at incipient failures.
The repair and restoration times result to be strongly affected by the presence of redundant or stand-by pumps. Unfortunately, VTT comments, no information is given by the CEDB on system configuration.

In addition to the classical statistical analyses, VTT performs also Bayesian analyses; in the latter, times between failures and restoration times are assumed to be exponentially distributed and the uncertainty on the parameter is described by a gamma distribution. As prior parameters, a shape parameter equal to 0.5 and a scale parameter much less than 1 are assumed; they correspond to a non-informative prior. The results obtained by the classical approach and the Bayesian one are quite comparable; they disagree of a factor less than 3 in most of the cases.We note that, as the results of (6) show, this factor represents also, in the classical approach, the disagreement between the estimates based on the assumption of the exponential distribution and the estimates based on the assumption of the distribution which

maximizes the likehood function. This is due to the non informative prior assumed for the Bayesian estimation.
An unavailability study is also carried out in the Bayesian framework. It is shown that the contribution of incipient failures to unavailability represents about 90% of the total unavailability.

3.3 NUKEM

NUKEM decides not to investigate on the quality of the data. The authors of (7) think it is difficult to judge on this matter without access to the data collection source. Their analyses are therefore based "only and completely on the information contained in the data".

As a first step for data grouping, 20 sets of pumps, homogeneous from the engineering point of view, are identified. The pumps of an homogeneous set have the same engineering and operating attributes, application type included. Afterwards, 9 sets at a higher level are identified. They are obtained by grouping the pumps pertaining to plants of the same type (PWR, BWR, conventional) and having the same application type (extraction, feed, booster). The 20 homogeneous groups are then subsets of the higher level sets.

NUKEM analyses the failure intensity of the component set, an approach appropriate to deal with systems with repairable components. For a repairable system, the failure intensity I(t) at time t is estimated as the number of failures occurred in the system in the time interval (t,t+h) divided by the product of the increment h and the number of components in use at time t. If I(t) is constant with time, it can be assumed that successive times-to-failure are independent, identically distributed exponential stochastic variables.
By graphic methods NUKEM shows that I(t) is approximately constant with time for most of the homogeneous subsets of pumps, whereas it is decreasing with time for all the composite sets. This is the result of the combination into one set of several subsets, characterized by different failure intensities and different operating periods; it does not correspond necessarily to a real effect.

For the estimation of times to failure, to repair and to restore, NUKEM considers as probability distributions the exponential and the Weibull ones. The method used for the estimation of the parameters of the two distributions is the maximum likelihood method (MLE). The goodness of fit is checked by the Chi-square test and the Kolmogorov-Smirnov test.
As to the time-to-failure (all failures), the fit of the exponential distribution is acceptable for 60% of the homogeneous sub-sets; the percentage of "good fits" reduces to about 25% for the composite sets, thus indicating the effect of inhomogeneities. 87% of the Weibull shape parameters are less than one, thus indicating failure rates decreasing with time. The fact that even the most homogeneous subsets cannot completely be described by exponential distributions is highlighted.
Though the exponential fit cannot be consided always good, with the aim of comparing the various groups of pumps NUKEM considers the mean times to failure (MTTF's)-all failures- as results of the exponential MLE. The MTTF's are plotted, together with their confidence intervals, for all the subsets and sets. It is shown that these MTTF's scatter by two orders of

magnitude (from 10^3 to about 10^5 h), much more than expected from their 90% confidence intervals. This is an indication of the strong differences existing between groups of pumps as to their reliability features.
As to times to repair, we note that their mean estimated value (all failures) varies from 10 h to 150 h for the pump subsets and is about 60 h for the whole pump set. As to times to restore, their mean estimated value varies from 10 h to 500 h for the pumps subsets and is about 270 h for the whole pump set. Thus, on the average, the mean restoration times are higher than the mean repair times by a factor of 4.

3.4 SINTEF

SINTEF divides the data into strata; each power plant is one stratum (8). Moreover, the data are grouped according to plant type (i.e. PWR, BWR, conventional). Non-parametric analyses of the data (for all failure modes) are performed, by using the following methods:

Kaplan-Meier plots: estimation of survival probability, using Kaplan-Meier's estimator

Hazards plots: estimation of the cumulative hazard, using the Nelson-Aalen estimator.

Fitted curves for the exponential and Weibull distributions are also drawn in the plots. These plots show that the exponential distribution does not agree with most of the data sets considered. However, the Weibull distribution fits reasonably well to all the data sets; the estimated shape parameter varies between approximately 0.5 and 1. Assuming a Weibull distribution, the maximum likelihood estimates of the MTTF's for all PWR, BWR and conventional plants are about 7500 h, 11500 h and 13000 h respectively; the corresponding shape factors are 0.67, 0.50 and 0.79.

3.5 JRC

The JRC analyst (9) identifies 21 groups of pumps, on the basis of engineering judgement. Each group can be considered homogeneous from the engineering point of view, i.e. it consists of components alike in design and application type. Then he looks for outliers in each group, i.e. for those pumps with a too high failure probabilities (f.p.) when compared with the remainder of the group. As a total, 5 outliers are identified: each of them has a f.p. which deviates of more than 10 standard deviations from the mean of its group. These outliers, if not put aside to be submitted to a separate treatment, would alter too much the statistical properties of the pump set.
As failure modes, "all failure types" and "complete failures", both in operation and on demand, are considered.
A failure rate trend analysis shows that a constant failure process can be accepted at the level of each of the above-mentioned groups; this holds for the pumps with at least 5 years of operating time.
For the estimation of component failure rate or repair rate (assumed to be constant with time) and failure probability on demand (assumed to be independent of the number of demands) he uses a unified model. He assumes a binomial process for failure or repair events and a beta distribution for failure or repair rate and failure probability on demand; complete renewal

after repair is also assumed. Binomial sampling is assumed for both processes on demand and times processes. The argument is that failures and repair times are recorded by whole time units (hours and minutes respectively) and thus they can be regarded as the outcomes of a Bernoulli trial in which each time unit is identified with a trial.
He estimates the component failure and repair parameters by a Bayesian method. He derives the same parameters at the pump group level, by performing the weighted arithmetic average of the parameters of the pumps forming the group. The weight is the number of years of observation of the component.

3.6 ENEA VEL

ENEA VEL divides the data into strata by Correspondence Analysis (CA)(10). The numerical plant identifier, the plant type and some pump engineering and operating attributes are used as variables in the CA. In a first application of CA the variable pump application is considered as "active" (i.e. directly contributing to the factorial analysis). In a second application of CA it is considered as "illustrative" (i.e. not directly contributing to the factorial analysis). The two resulting stratifications are quite unlike each other. Of the four strata identified by each stratification, one or two contain only 3 or 4 components; they are too small to be analysed . We note also that the statistically homogeneous groups of pumps identified by CA are quite different from the groups identified, on the basis of engineering judgement, by the other participants. ENEA VEL recognizes that these results are of doubtful usefulness In their opinion, CA can give usefull results, provided that a strong support of the engineer to identify all the influencing variables and the relative importance is available.

For the estimation of failure rate, the renewal model "component as bad as it was" is considered, in addition to the usual one "component as good as new". In this case times to failure are all counted from the beginning of the observation; no maintenance effect on lifetimes is considered.
A trend analysis of the failure rate is performed for the identified pump groups. The following cases are considered: all the failures, the failures occurred in different operating time windows to detect "infant mortality" or ageing effects, all the failures with the exception of those due to errors in design, or manufacturing, or installation. The failure rate shows no clear trend in these cases; nevertheless, most of the pump groups gets an increasing failure rate after the first 10000 operated hours.

ENEA VEL notes that the usual methods for the estimation of component failure probability do not exploit all the information that the CEDB makes available. For each identified stratum, these methods consider the basic failure-data, but do not take into account at all the repair data associated to failure data. They do not consider repair data such as the description of the parts failed and consequently replaced, the failure mecanism and the failure causes. As a new approach to the estimation of the component failure probability, ENEA VEL considers the component as a system (usually a series system) and performs its logical breakdown into parts, following a fault tree technique. The failure probability of the component can thus be obtained as a function of the failure probabilities of its constituent parts, the failure of which are regarded as initial events. The problem so transforms into the estimation of the failure probability of each constituent part on the basis of the CEDB failure/repair data. It is assumed

as lifetime of a constituent part the operating time between two successive replacements of this part, i.e. the hours operated between two successive failures in which this part is recorded as failed. It is recognized that this evaluated lifetime is not correct , as the CEDB does not collect data on preventive maintenance; i.e. all the planned replacements are not recorded and this evaluation does not take them into account. Furthermore, we add, the CEDB does not specify if a part recorded as failed in a component failure-event failed spontaneously (i.e. the event corresponds to a genuine failure of the part), or had an induced failure (i.e. the event corresponds to a suspension of the observation of the part).
Moreover, ENEA VEL has developed some failure models that describe degradation phenomena, which may affect the mechanical parts of a component, such as corrosion, erosion, fatigue, and errors of the operator during maintenance. The probabilistic nature of the failure of a part due to a certain phenomenon derives from the statistical variations of the variables of the physical laws governing the phenomenon. For instance, in the stress-strength model the failure probability is a function of the statistical distributions of the load applied on the part and of the part strength. By these so-called "physical failure-models", ENEA VEL can predict the failure probability of a part of a component as a result of a well defined failure process as described by CEDB data.
An example of breakdown of a centrifugal pump into parts is given in (10). For a few parts of the pump, the failure probability is estimated both by the usual statistical processing of the CEDB data related to the above-mentioned homogeneous groups of pumps and by the application of the "physics models" to the same data. The results are in fairly good agreement.

3.7 EDF

The aim of the work of EDF (11) was the selection of a set of failure data suitable for the application, by ENEA TIB AQ Roma, of competing risk models for the estimation of the reliability at the component level and at the system level, in a rigorously Bayesian framework. C.A.Clarotti (ENEA) intended to demonstrate that the method to be used for the estimation varies according to the objective, in this case according to whether the estimation of the reliability has to be made either at the component level or at the system level. This application has not been performed, as ENEA TIB did not continue to participate in the BE.

As first task performed by EDF, (11) describes the selection of a set of pumps which can be considered "exchangeable", i.e. with identical design attributes and as similar as possible operating conditions. A group of 30 extraction pumps of similar PWR plants is chosen.
A screening of the failures associated to this selected group of pumps is thus carried out. For the purposes of this work, a failure is defined by EDF as an event characterized by the (immediate or deferred) inability of the pump to perform its function when requested (i.e. with plant in operating condition). As a consequence, those events which did not cause component unavailability during plant operation arc not considered as true failures; minor incipient failures, without any consequence for the operation, and potential failures, i.e. anomalies detected during maintenance, are to be discarded. Two groups of (true) failures are then identified: the catastrophic failures, characterized by a sudden and complete loss of function and immediate unavailability, and the incipient ones, i.e. anomalies characterized by a delayed unavailability for repair.

It is to be noted that the EDF analyst needed knowing the plant condition during repair to perform the work described above. Unlikely CEDB does not give this information item. He used some tables giving plant conditions as a function of the calendar time for all the power plants considered in the BE. The coordinators distributed these tables to all participants during the second phase of the BE. Only the EDF analyst made use of them.

4. Major findings and lessons learnt

4.1 Objective assumed for the analysis

The BE has shown the strong relationship existing between the objective assumed by the analyst and its approach to data interpretation and analysis. INTERATOM, for instance, is interested in demonstrating the difficulties of deriving reliable parameters for PSA from raw data extracted from a "multi-purpose data bank" such as CEDB. It concentrates its attention in identifying a suitable set of pumps and the associated relevant set of failures, in checking the related data quality and consistency, whereas it uses a very simple method for the failure rate estimation. VTT, which is more interested in comparing methods, focusses its effort in exploratory data analysis and in testing different approaches for estimation.

4.2 Data interpretation and data quality check

Data understanding has been a difficult task for the participants having no specific knowledge of the CEDB data bank. The structure of CEDB data and the relative coding are complex. According to some participants (INTERATOM, VTT), the definition of some codes related to failure mode is not clear.
It is noted by INTERATOM, for instance, that some codes of failure mode are not "exclusive", i.e. they do not univocally identify the characteristics.of one failure type. As a matter of fact these codes can refer to both actual failures, i.e. failures occurred during component operation, and potential failures, i.e. anomalies discovered during the preventive maintenance of the component in out-of-service condition and judged capable of impairing the function if not corrected. We remark that, if the analyst is interested in failures actually occurred, he has to previously select failures on the basis of the code "failure detected during (component) operation", so as to discard failures detected during maintenance.
We conclude by saying that a coordination meeting devoted to data interpretation would have been a great help. The analyses should have had to be started by the participants only after having obtained a consensus on this first step. In this regard, we note that no intermediate meeting could be held during the BE due to lack of funds. During both phases of the BE, it was not possible to organize coordination meetings to compare partial results and agree on how to proceed for the following step.
As to the often unsatisfactory quality of the data, we agree with INTERATOM that an effort should be made to improve the situation. This is a major problem of all the component data banks which store data collected by the operating staff, i.e. persons having no background in reliability and data analysis.
How to improve data quality is an area of research of some Organizations managing important national data banks. VTT suggests to make available to the operating staff the output of a

simple descriptive data analysis, showing graphical representations of very understandable indicators of performance of the components for which they collect data (5). In the VTT opinion, this would be of help to plant operator to monitor equipment performance, to organize maintenance, etc. We think that also data quality would derive benefit from such an initiative, somehow involving the operator in the use of the data. Becoming a data user would make him aware of the importance of collecting good data.
As regards this matter, we recognize that the comments made by INTERATOM in their report at the end of the first phase of the BE have been of great help to JRC staff for the revision of the data set to be used in the second phase.
A few participants (e.g. the JRC analyst (9)) defined some pumps as outliers (i.e. with a too high failure probability) and put them aside, to be considered separately. INTERATOM discarded some clearly dependent failures from the sample considered. Again, we repeat, a coordination meeting would have been very useful to deal with all these questions.

The difficulties in data interpretation were enhanced by the censoring of the data previously performed by the JRC, a censoring made necessary to guarantee the anonimity of the sources. The system functional flow sheets, made available to the participants together with the data set for the second phase of the BE, was of great help, mainly to those who analysed repair and restoration times and made unavailability assessments (VTT).

4.3 Data stratification

All the participants agree on the necessity to use firstly engineering judgement to identify the most usefull variables, before using statistical analysis tools. For instance, ENEA VEL recognizes that the results obtained by the application of Correspondence Analysis for data stratification are of uncertain usefulness. A discussion among the participants on criteria and methods used for data stratification and on the results otained would have been very beneficial. A consensus on some common sets of components to be considered for the following steps of the analysis would also have had to be gained.

4.4 Estimation

All the participants agree on the fact that the usual renewal model "component as good as new (after repair)" is not realistic. We can say the same for the model "component as bad as it was" considered by ENEA VEL , as it ignores any maintenance effect on lifetimes. Renewal model is still a major problem in the area of reliability estimation for repairable components.
The suggestion of ENEA VEL to consider the component as a series system of piece-parts and to concentrate on the estimation of the reliability of these parts looks very attractive. Nevertheless, this would imply a more careful monitoring of the component parts, i.e. a correct diagnosis of the part which caused the failure and the recording of all the part replacements, i.e. occurred during both preventive and corrective maintenance. This would call for an improvement of the CEDB data collection scheme and an increase of the effort of the data collector.
We note that most of the estimations made by the participants give as results failure rates decreasing with time. This is probably due to the superposition of many effects. INTERATOM justifies this as follows:

- most of the operating histories refers to the first years of the components life. Many events can still be framed in an infant mortality phase. This, combined with the effect of a learning by the operating staff, would explain a real reliability growth trend;
- the overall reporting activity in some plants is decreasing in time (the data collector tends to ignore events of minor importance);
- mixing components or sets of components, each one with approximately constant failure rate, leads to a common decreasing failure rate.

4.5 Suggestions of CEDB improvements

Some participants (mainly VTT and INTERATOM) have highlighted the necessity of, or the advantages offered by, some improvements of the CEDB data collection scheme. We summarize these suggestions in the following.

The repair should be better described. Among the component "parts failed", the part which failed as the first and has to be regarded as the immediate cause of the component failure, should be identified. The condition of the plant during repair, strongly infuencing repair duration, is not specified. For some safety-related components, the knowledge of the prescriptions of the plant technical specifications as to the maximum allowable outage time for repair would be of help to the analyst for the interpretation of some unusual restoration times.

The multiple field "related failures" of the failure description, recording the utility failure codes of linked events, had been censored by the JRC staff to guarantee source anonimity. The information on this linkage between failures is important; it has to be expressed by a different coding and made available to all users.

The results of VTT analyses have shown that system layout (number of the trains, capacity and operating mode of each train) strongly affects restoration and repair times and availability. This information item, which is of use also for data interpretion, is not given by CEDB. The data collection scheme should be revised to allow its recording among the operating characteristics.

Preventive maintenance should also be recorded, at a level of detail similar to that adopted for the corrective maintenance. According VTT, this would allow the assessment of the overall performance of the component (reliability, availability, maintainability). Furthermore, as already said, it would allow a better modelling of the component renewal due to maintenance (ref. to ENEA VEL approach).

4.6 General comments on the results obtained

A comparison between the results of the analyses performed by the various participants is very difficult. This is due to the combined effect of several factors, which we have tried to identify and deal with in the previous paragraphs. The participants assumed different objectives for their analyses, had difficulties in data understanding, adopted different criteria and methods for data stratification. All this led them to analize different data. No group of pump, for instance, was examined by all participants. Nevertheless, a few participants sometimes chose the same set of pumps for analysis; we note that this does not imply that they considered the same set of failures. It turns out that, in most cases, the estimates they obtained are quite comparable, i.e. disagree of a factor of less than three. Only sometimes these estimates differ of up to one order of magnitude (mainly in the case of sudden failures, i.e. of samples of few events). To understand the reason of this, further investigation would be necessary; in particular, a through study of the estimation methods used by the participants would be useful. Probably this study could not be made only on the basis of the reports produced by the participants for the BE.

5. Summary and conclusions

A EuReDatA Benchmark Exercise on data analysis was organized and coordinated by the JRC. Aim of the BE was the comparison of the methods used by the participants for the estimation of component reliability from raw data. As reference data set, CEDB raw data related to pumps were used.
A description of the approach adopted by each participant has been given. The major findings of the BE and lessons learnt have then been identified and commented.
A comparison between the results of the analyses performed by the various participants is very difficult. This is mainly due to the fact that the participants adopted different criteria for the choice of the sets of pumps to analyse and almost always analysed different data.sets Also the impossibility, due to lack of funds, of organizing intermediate meetings to compare partial results had some effect on that.
Nevertheless a few participants sometimes examined the same group of pumps. The estimates they obtained are often quite similar; in the case of estimates based on samples of few events (sudden failures), these estimates can disagree of up one order of magnitude. Further effort would be necessary to fully interpret that.

This BE on data analysis has been the first initiative of this kind taken by EuReDatA. Analyses of great interest have been made by the participants and very interesting insights have been gained .We think that all of us, involved in the BE, have learnt very much and that this BE, as our first experience, has been a great success.

Aknowledgements

S.P. Arsenis of JRC Ispra is thankfully aknowledged for the fruitful discussions and suggestions received.

References

1) Balestreri, S. and Carlesso, S. (1990) "The CEDB Data Bank; informatic structure and use", proceedings of the Eurocourse on Reliability Data Collection and Analysis, CEC JRC Ispra, October 8-12, 1990, Kluwer Academic Publishers.

2) Besi, A. and Colombo, A.G. (1989) "Report on the on-going EuReDatA Benchmark Exercise on Data Analysis", proceedings of the Sixth EuReDatA Conference on Reliability Data Collection and Use in Risk and Availability Assessment, Siena, Italy, March 15-17, 1989, Springer-Verlag, 253-361.

3) Besi, A. and Colombo, A.G. editors (1990) preprints of the proceedings of the conclusive Workshop of the EuReDatA Benchmark Exercise on Reliability Data Analysis, CEC JRC Ispra, April 5, 1990.

4) Pamme, H. (1990) "Derivation of reliability parameters from a Component Event Data Bank", INTERATOM, preprints of the proceedings of the conclusive Workshop of the EuReDatA Benchmark Exercise on Reliability Data Analysis, CEC JRC Ispra, April 5, 1990.

5) Simola, K., Huovinen, T., Komsi, M., Lehtinen, E., Lyytikäinen, A. and Pulkkinen, U. (1989) "VTT's contribution to the EuReDatA Benchmark Exercise on Data Analysis; preliminary analyses results", Technical Research Centre of Finland (VTT), presented at the meeting of the participants in the BE, Siena, March 13, 1989.

6) Simola, K. and Pulkkinen, U. (1990) "EuReDatA Benchmark Exercise on Data Analysis; VTT's final reports", Technical Research Centre of Finland (VTT), preprints of the proceedings of the conclusive Workshop of the EuReDatA Benchmark Exercise on Data Analysis, CEC JRC Ispra, April 5, 1990.

7) Leicht, R. and Wingender, H.I. (1990) "EuReDatA Benchmark Exercise on Data Analysis; Report prepared for the Workshop on Reliability Data Analysis", NUKEM GmbH, ibidem.

8) Lydersen, S. and Samset, O. (1989) "EuReDatA Benchmark Exercise on Data Analysis; preliminary results from SINTEF" presented at the meeting of the participants in the BE, Siena, March 13, 1989.

9) Jaarsma, R.J. (1990) "EuReDatA Benchmark Exercise on Data Analysis; final report", preprints of the proceedings of the conclusive Worshop of the EuReDatA Benchmark Exercise on Reliability Data Analysis, CEC JRC Ispra, April 5, 1990.

10) Righini, R. and Zappellini, G. (1990) "EuReDatA Benchmark Exercise on Data Analysis", ENEA VEL, ibidem.

11) Piepszownik, L. (1990) "EuReDatA Benchmark Exercise; engineering analysis of the pump sample", EDF, Direction des Etudes et Récherches, ibidem

DEMONSTRATION OF FAILURE DATA BANK, FAILURE DATA ANALYSIS, RELIABILITY PARAMETER BANK AND DATA RETRIEVAL

R. LEICHT, H.J. WINGENDER
NUKEM GmbH
P.O. Box 1313
D-8755 Alzenau
FRG

1. Introduction

1.1 Overview

This paper describes the content of the demonstration which is intended to run directly through a personal computer in order to show the operation of the codes. A compiled version of the demonstration will be made available to the participants of the course. In order to perform the demonstration in time, a few typical example cases are selected and treated with simplified versions of the codes.

The demonstration begins with the Failure Data Bank code (FDB) and includes two examples of data sets: one from the EuReDatA benchmark exercise and a second example with some sets of failure data from car dashboard instruments and break linings.

In a second step these data are processed by the Failure Data Analysis code (FDA) using Exponential, 2- and 3-parameter Weibull lifetime distributions and appropriate checks of the goodness of the fits.

The resulting reliability parameters are transferred to the Reliability Data Bank (RDB) from which they can be retrieved for reliability assessment purposes. This will be demonstrated by means of the Fault Tree code (FTL).

1.2 Structure of the Code System

The requirements of reliability and risk analyses have initiated the in-house development of a variety of computer codes which are tailor-made for the specific needs of the analyst. The need for an integrated software system comprising all these codes and data was soon recognized.

J. Flamm and T. Luisi (eds.), Reliability Data Collection and Analysis, 299–325.

It was obvious that such a system should provide

- interactive or dialogue use,
- easy and fast access to desired information,
- low hard and software costs, and
- ability for an optimal design of the user-surface.

All these requirements are fully met by the modern Personal Computer (AT), which was therefore chosen as hardware basis of the system. Today, the integrated system called CARARA (Computer Aided Reliability And Risk Assessment) is realized to a great extent, and is continuously being further developed [1]. The structure of this system and the modules forming it are described in the following.

The structure of the CARARA system is graphically represented in fig. 1. The system consists of interlinked data banks and programs which are required to either assess the reliability or the release risk of a technical system or plant. The basis of all reliability work is the failure behaviour of components as recorded from either field or test experience. These data are organized and managed in the component/failure event data bank FDB. Homogeneous component subsets are identified, and the life times data file is prepared for the statistical analysis with the program FDA.

The results of this analysis - or reliability parameters from external sources - are stored in the reliability data bank RDB and are available for use in fault tree or risk analyses.

The reliability of technical systems which are composed of a variety of components is quantitatively evaluated with the fault tree code FTL. Input data required are information on the structure of the system to be analysed and reliability data of the basic events of the tree. The fault tree code is used as an example for data retrieval. It is not described here in detail since this would exceed the frame of the course.

The release risk, i.e. the risk of an accidental release of toxic or radioactive material into the environment, is assessed with the STAR code, and physical/mathematical models describing the transport of this material must be provided in addition to system structure (fault trees) and reliability data. This section of the code system is neither used nor described in the demonstration.

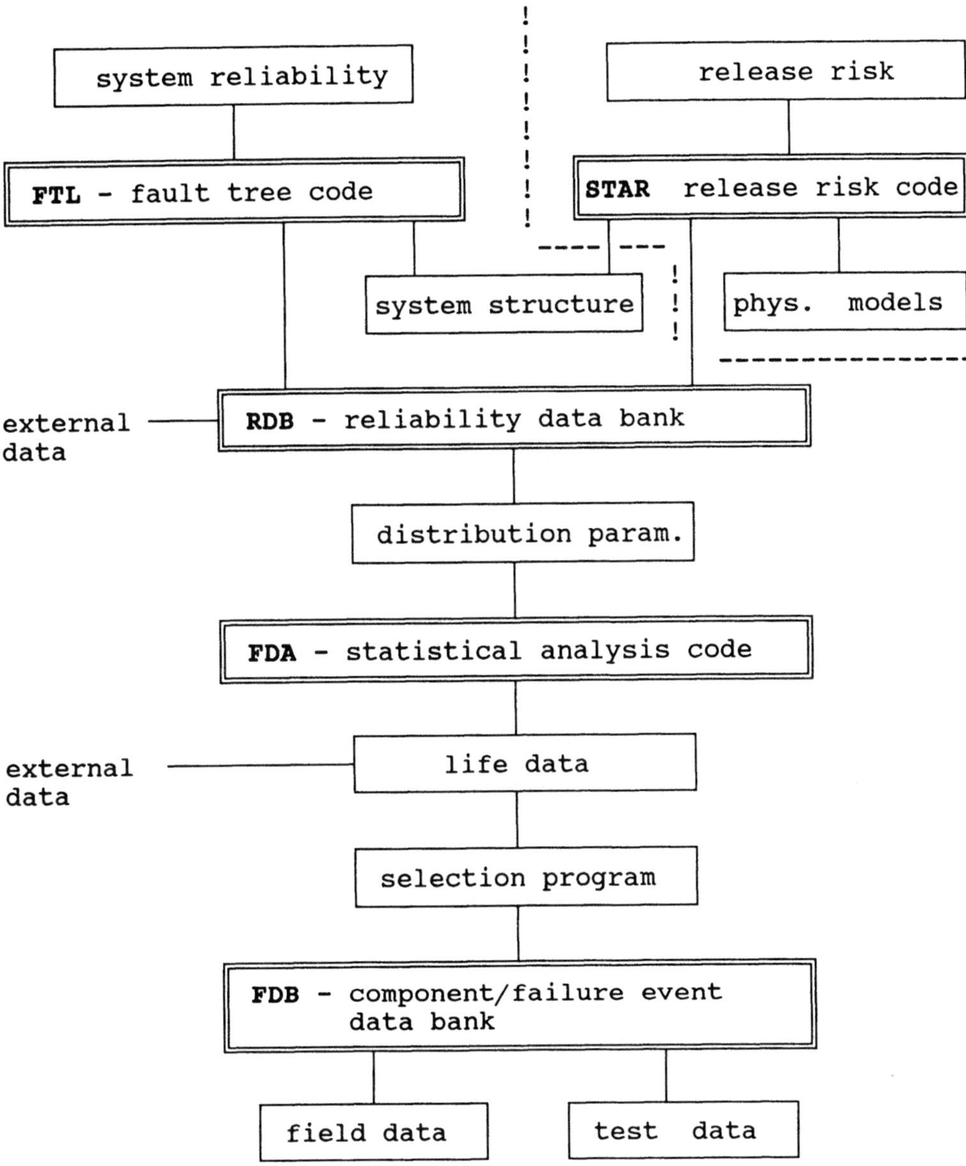

Figure 1
Schematic representation of the CARARA system structure

2. FDB - Data Bank for Component/Failure Event Data

2.1 The FDB Structure

As a first step for the statistical evaluation of reliability parameters of technical components, data concerning such components and their life histories are to be collected and stored in an appropriate data bank.

Data collection is in general performed within a plant where a variety of technical components are operating, many of them identical, distinguishable only with respect to their location in a system or in the plant.

Prerequisites for data collection are the ability to identify homogeneous subsets of components, i.e. sets of similar components with similar failure behaviour, and also the ability to evaluate the life history (periods of operation, failures, and repairs) of each individual component.

According to the rules of relational data bank design, these requirements are met by introducing a system of 3 related data base files:

(1) Component Type File
component type specific information, such as design parameters and specifications, manufacturer, etc.
Key: component type number

(2) Component File
individual component specific information, such as system, location, serial number, operating parameters, etc.
Key: component identifier

(3) Failure Event File
operating history of individual components
failure event, total operating time and/or number of cycles/demands at time of failure, repair time, etc.
Key: failure event number

The resulting overall structure (entity relationship model) of the FDB data bank system is illustrated in fig. 2, and a more detailed structure of the data bank files is given in fig. 3.

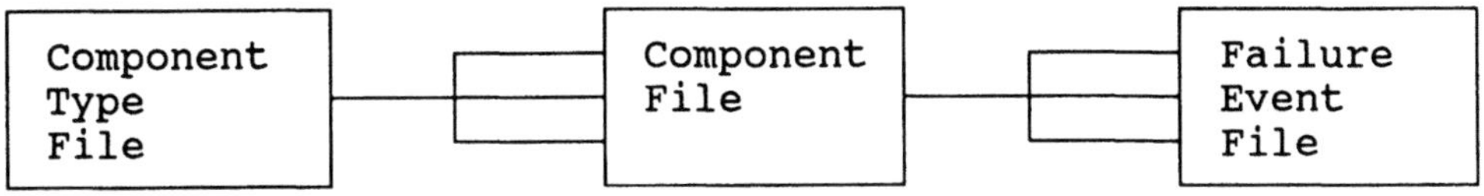

Relations: 1 : N 1 : N

Figure 2
Entity Relationship Model of the FDB system

In general, two failure modes are to be considered:

- failure during operation, and
- failure on demand.

For the evaluation of reliability parameters the observation period, the total operating time and the number of demands or operating cycles is recorded for every component. Since the components are assumed to be repairable, more than one failure (and different failure modes) can occur during the observation period.

Appropriate selection programs are used for identifying homogeneous component subsets - a task which cannot be performed without engineering judgement and experience. The conditions for the selection can be stored in a selection criteria file.

The application program TIMES extracts from these homogeneous subsets the life data which are required as input data for the statistical analysis with FDA.

Depending on the results of the statistical analysis, it may be necessary to iterate the steps of identifying homogeneous subsets and analysis until criteria for homogeneity (e.g. applicability of exponential life distributions) are fulfilled.

TIMES also provides the probability of failures on demand including the corresponding uncertainties, e.g. the 5% and 95% confidence limits of the failure probabilities. These data can be transferred to the reliability data bank RDB without being further analysed.

F D B - Failure Data Bank

Component Type File

o key: component type number
o manufacturer
o design specifications
o classification/licensing
o ...

Component File

o key: component identifier
o component type number
o serial number
o installation (system, location)
o operating specifications
o environmental conditions
o modes of operation
o maintenance type
o dates of construction/initial operation
o ...

Failure Event File

o key: failure event number
o component identifier
o date/time of failure
o operating hours/calendar time at failure
o repair/unavailability time
o failure detection
o failure mode
o failure description
o failure causes
o failed parts
o failure consequences
o measures taken
o ...

Figure 3
Fine structure of the FBD system data files

2.2 The Data Sets

2.2.1 EuReDatA Benchmark Exercise Data Set

A Benchmark Exercise on data analysis was initiated in 1987 by EuReDatA with the aim to compare the methods used by the various participating organizations. The CEC-JRC Ispra has been charged with the coordination. The results of this exercise are compiled in [2].

The reference data were raw data taken from the Component Event Data Bank, managed by the JRC Ispra [3]. For the second phase of the BE, a data set concerning 114 pumps of the power plant condensate and feedwater system F08 with a total of 440 failures has been provided.

The raw data set obtained from JRC-Ispra contained information in coded matrix format. The coded matrix format files have been used for easier storage of these data in a relational data base as required by the FDB system. Each record of the matrix format files included a description of component and operational conditions (fixed length), and a description of the failure events (if any) observed for this component (variable length, up to 20 failures).

For optimal storage, this matrix format has been partitioned into two relational data files, one containing component related information (PUMP), the other the particular failure events (FAIL). As an unambiguous external key, the component number was used to relate these two data bases.

Homogeneous component sets have to be identified in order to allow a significant statistical analysis of the failure data. The criteria to be considered when selecting such sets are system/operation related criteria and failure/event related criteria. They are discussed in the following.

Failure/Event Criteria:

The most important criterion for defining a failure event is the failure mode. A pump can fail either on demand, i.e. when starting its operation is intended, or during operation, i.e. the pump has already been in operation for a certain time when the failure occurs.

The type of reliability parameters to be derived from the data depends on the failure mode: the failure on demand is characterized by a (time independent) failure probability whilst the failure on operation is described by a failure rate.

Other important keys concerning failures on operation are the suddenness of the failure and its degree of seriousness. An overview of the combinations and their frequency in the database is given in table 1.

failures on operation	sudden	no output	13		
		outside specs.	79		
		total		92	
	incipient	no output	0		
		outside specs.	332		
		total		332	
	total				424
failures on demand		fails to start	14		
		outside specs.	2		
		total			16
total					440

Table 1
Overview of failure modes and characteristics derived from the component data file PUMP

System/Operation Criteria:

System and operation specific information is taken from the flow diagrams provided by JRC for the condensate and feed-water system pumps F08.

As a first step, we tried to identify homogeneous sets of pumps on a "microscopic" level: pumps are combined to sub-sets when they are of the same type, have the same design characteristics and work under the same operating conditions. With this approach, the highest obtainable degree of homogeneity should be reached.

20 types of pumps have so been identified and chosen as homogeneous subsets for the analysis. They are compiled in table 2 with their relevant design or operating parameters.

Sub-set No.	Plant No.	Plant Type	Pump Type	Design Power [kW]	Oper. Flow [m³/s]	Oper. Press [bar]	Oper. Head [bar]	Oper Temp [°C]	Numb of Pmps	Numb of Fail
01	1-5	PWR	Extr.	2240	0.490	32.0	31.9	32	18	55
02	1-5	PWR	Feed	3580	0.810	63.0	31.0	180	14	95
03	6	BWR	Extr.	1365	0.660	15.6	15.5	33	3	0
04	6	BWR	Boost	2170	0.660	39.0	27.0	34	3	3
05	6	BWR	Feed	4635	0.800	70.0	36.0	192	3	17
06	6	BWR	Feed	15	0.017	5.9	5.6	33	4	4
07	7-10	CONV	Extr.	880	0.214	31.4	31.4	35	8	35
08	7-10	CONV	Boost	100	0.157	15.7	7.8	162	12	44
09	7-10	CONV	Feed	4217	0.157	220.0	204.0	168	12	80
10	11	BWR	Boost	294	0.357	9.0	-	29	2	1
(11)	11	BWR		294	0.357	20.0	-	57	2	-
12	11	BWR	Feed	442	0.037	76.5	-	135	3	21
13	12	PWR	Extr.	1200	0.238	30.0	30.0	34	6	5
14	12	PWR	Feed	3500	0.385	100.0	63.0	115	6	12
15	13	PWR	Extr.	3125	0.520	47.0	46.0	40	3	9
16	13	PWR	Feed	4909	0.856	73.5	40.5	183	3	14
17	14	PWR	Boost	1550	1.000	23.0	13.0	180	3	8
18	14	PWR	Feed	7270	1.000	77.0	54.0	180	3	1
19	16	PWR	Boost	445	0.406	18.1	8.5	250	3	9
20	16	PWR	Feed	3600	0.406	72.0		250	3	7

Table 2
Identical pump subsets and the related operating parameters and frequencies

2.2.2 Brake Linings Data Set

An important non-nuclear field with increasing need for reliability engineering is the car manufacturing industry. The market forces the car manufacturers to extend their warranty periods, and thus the suppliers of manufactured components have to demonstrate and to guarantee specific reliability requirements.

The German Association of Automotive Industry (VDA) has edited a series of publications dealing with quality and reliability control and recommending methods and procedures for failure data analysis [4]. An example of failure data given there is taken as a data set to be analyzed in the demonstration.

This data set describes a sample of front brake linings and later illustrates the need of a 3-parameter Weibull life time distribution with a failure-free "time" (i.e. mileage). The total number of observations is 65. The mileages at failure is recorded for 24 parts. "Survived" mileages of 41 brake linings are also given, i.e. linings which were still intact at the end of the observation period.

2.2.3 Dashboard Instrument Data Set

Another example from the automobile supply industry concerns dashboard instruments [5]. A special warranty data bank stores information of the monthly production number of a certain device. As failures of this equipment occur after starting operation, i.e. after delivery of the car, the resulting repairs or exchanges of defective devices are recorded and assigned to the particular production periods.

In this way, data files are compiled with a typical format shown in table 3. From these data files, life times and censoring times are derived for statistical Weibull analysis. This data example illustrates that in some cases systematic influences occur which cause non-stochastic distortion of the data, as shown in the next paragraph.

CLIENT 7 COMP.NO. : 120 108 -1 -1

PRODUCTION PERIOD : 8601 TO 8809 LIFE TIME IN MO DATE: 25.10.88

production year	production month	number of failed components: month 1 to 12 of operation / month 13 to 24 of operation / month 25 to 36 of operation												production number
86	1	0	6	4	8	16	3	1	5	6	2	2	1	19486
		4	6	1	8	4	1	1	4	2	4	1	2	
		0	1	0	0	0	0	0	0	0	0	0	0	
86	2	3	5	8	24	14	4	3	14	9	11	7	9	35562
		9	10	11	6	0	0	4	1	2	1	0	0	
		3	0	0	1	0	0	0	0	0	0	0	0	
86	3	1	5	15	18	4	8	13	24	11	17	9	18	33193
		11	13	12	0	4	5	2	3	3	0	0	1	
		1	0	0	0	0	0	0	0	0	0	0	0	
86	4	1	8	15	4	13	12	9	19	22	10	19	13	33276
		23	13	4	9	5	2	1	2	2	0	4	0	
		0	0	0	0	0	0	0	0	0	0	0	0	
86	5	2	4	4	13	32	18	17	11	11	18	20	8	31508
		13	6	6	10	1	2	1	1	0	4	1	0	
		0	1	0	0	0	0	0	0	0	0	0	0	
86	6	0	4	7	10	18	15	17	18	17	16	7	17	38894
		1	8	14	14	1	1	0	3	5	1	1	0	
		0	0	0	0	0	0	0	0	0	0	0	0	

Table 3
Format of dashboard instrument warranty data [5]

3. FDA - Program for Statistical Analysis of Life Data

3.1 The Code FDA

The reliability parameters required for the quantification of the reliability of a system are provided by the statistical analysis of the life data of the technical components within that system.

A widely used method for reliability analysis is that of fault tree analysis. The failure behaviour of the basic events - mainly certain failures of components - is in general described by constant failure rates. This assumption, however, should be confirmed by a statistical analysis of life data. The object of such an analysis is to describe test or field life data by mathematical life distribution functions whose parameters are determined in a fitting procedure. Statistical tests then allow a decision as to whether a certain life distribution function is acceptable or not.

FDA is a computer code developed for this purpose, interactively running on a personal computer and also able to present the results graphically. The essential features of the present FDA version are described in the following. FDA can easily be modified for special tasks and is continuously being improved [6].

The input data required by FDA are life data of components and can be of the following types:

(1) operating times until the first failure
(2) operating times between subsequent failures (repairable components only)

(3) observation times with no failure occurred (survival time of a component)
(4) operating time after which an intact component has been replaced (preventative action)

(1) and (2) represent 'true' life times, while (3) and (4) are times of survival which are also called censoring times. Censoring times provide important information for those life distribution functions which correspond to time dependent failure rates. FDA accepts both life and censoring data as input data.

The aim of the statistical analysis is to describe the distribution of observed life data with mathematical distribution functions which are as simple as possible.

The life distribution function most important in reliability analysis is the exponential distribution which corresponds to a constant failure rate. This failure rate is also the only parameter of the exponential distribution.

Weibull distributions provide a very flexible approach to describe time dependent failure rates. The 2-parameter Weibull distribution is characterized by a location parameter and by a shape parameter which allows the approximation of various other distribution functions, e.g. the normal and the lognormal distributions. The Weibull distribution is equivalent to the exponential distribution when the shape parameter is set equal to 1.

In special cases where failures are expected or observed only after a certain 'failure free' time, the appropriate distribution is the 3-parameter Weibull distribution with this failure free time as third parameter. This can be of interest for components which fail mainly by (planned or predetermined) wear, such as brake linings.

The parameters of the distribution functions are determined so that the observed life data are reproduced as close as possible. In FDA, the fitting of the exponential and of the 2-parameter Weibull distribution is performed according to the maximum likelihood method. In the case of the 3-parameter Weibull distribution, the failure free time is determined by an additional least squares fit.

The goodness of the fit is checked in a Chi-Square test and according to Kolmogorow-Smirnov.

3.2 Results

The results of an analysis with FDA are the parameters of the 3 distribution functions and the goodness of fit criteria. In addition, graphical representations can be provided either on the monitor screen or on a HPGL-compatible plotter.

4 different graphical outputs are offered:

- cumulative probability distributions
- distribution histograms used in the least squares fit
- distribution functions
- hazard functions (failure rates)

All these representations include data points with (statistical) error bars which are derived from the observed data, thus enabling an immediate comparison between observation an fits.

Scaling is performed automatically, and the graphics can optionally be provided with linear or logarithmic axis or on a Weibull probability paper.

3.3 Examples

The first example is the analysis of the brake lining data [4]. FDA provides a text output file shown in table 4. Some general information is given in the upper part while the lower part shows the result of the fits with 3 different life time distributions. In this example, the "life time" is actually a mileage in 1000 km units.

T denotes the mean life time in the case of the exponential distributuion and the location parameter for the Weibull distributions. b is the Weibull shape parameter, indicating increasing failure rate for b>0, constant failure rate for b=1 and decreasing failure rate for b<1. Finally, t_0 denotes the failure free time, which in the present case amounts to 8830 km.

MaxDv is the maximum deviation of the cumulative fit curve from observed data and is required for the Kolmogorow-Smirnow goodness of fit test (probability PKol). This test is strictly applicable only for the exponential case and must be seen as an auxiliary indicator otherwise. A more general test is the Chi-square test. Chi square (X^2) and the corresponding probability $P(X^2)$ are given in the table.

EWA - Exponential and Weibull Analysis Version 25.02.90 C NUKEM GmbH SiT/Lc

Analysis of Life Time Data File: VDA

Number of Life Times	: 65	Minimum	:	1.00
Number of Failures	: 24	Maximum	:	105.00
Censorings	: 41	Mean	:	34.89
Life Time Classes	: 14	Sum	:	2.27E+03
early Failures	: 0	Po	:	0.00E+00

Distribution	Fit	T	b	to	MaxDv	PKol	X²	P(X²)	<T>
Exponential	MLE	9.45E+01			.1993	.000	5.18	.000	9.45E+01
Weibull	MLE	6.67E+01	2.17E+00		.0792	.200	2.85	.001	5.90E+01
Weibull	LSQ/MLE	5.87E+01	1.58E+00	8.83E+00	.1000	.200	2.26	.007	5.36E+01

Table 4
Results of the statistical life time analysis with FDA - front brake linings example [4]

Fig. 4 shows the cumulative distribution derived from the observed brake lining failure and survival data and 3 curves resulting from the 3 fit functions. As one can clearly see, the exponential function fails completely to meet the observed data, while the 2-parameter Weibull distribution is in a much better agreement with the data. The best fit, however, is obtained with the 3-parameter Weibull distribution, indicating the need of a failure free time resp. mileage.

The second example is taken from the EuReDatA Benchmark exercise [2] and illustrates two cases: In the first case, the failure behaviour of the selected component set can sufficiently be described by an exponential function (table 5, fig. 5). In the second case, a 2-parameter Weibull distribution is required (table 6, fig. 6), thus indicating time dependent - in the present case decreasing - failure rates. The reasons of these phenomena can be inhomogeinity of the sample or other systematic influences.

The evaluation code has been slightly modified in the second example. 3-parameter Weibull function is not considered, but upper and lower 90% confidence limits of the exponential mean life time are provided instead.

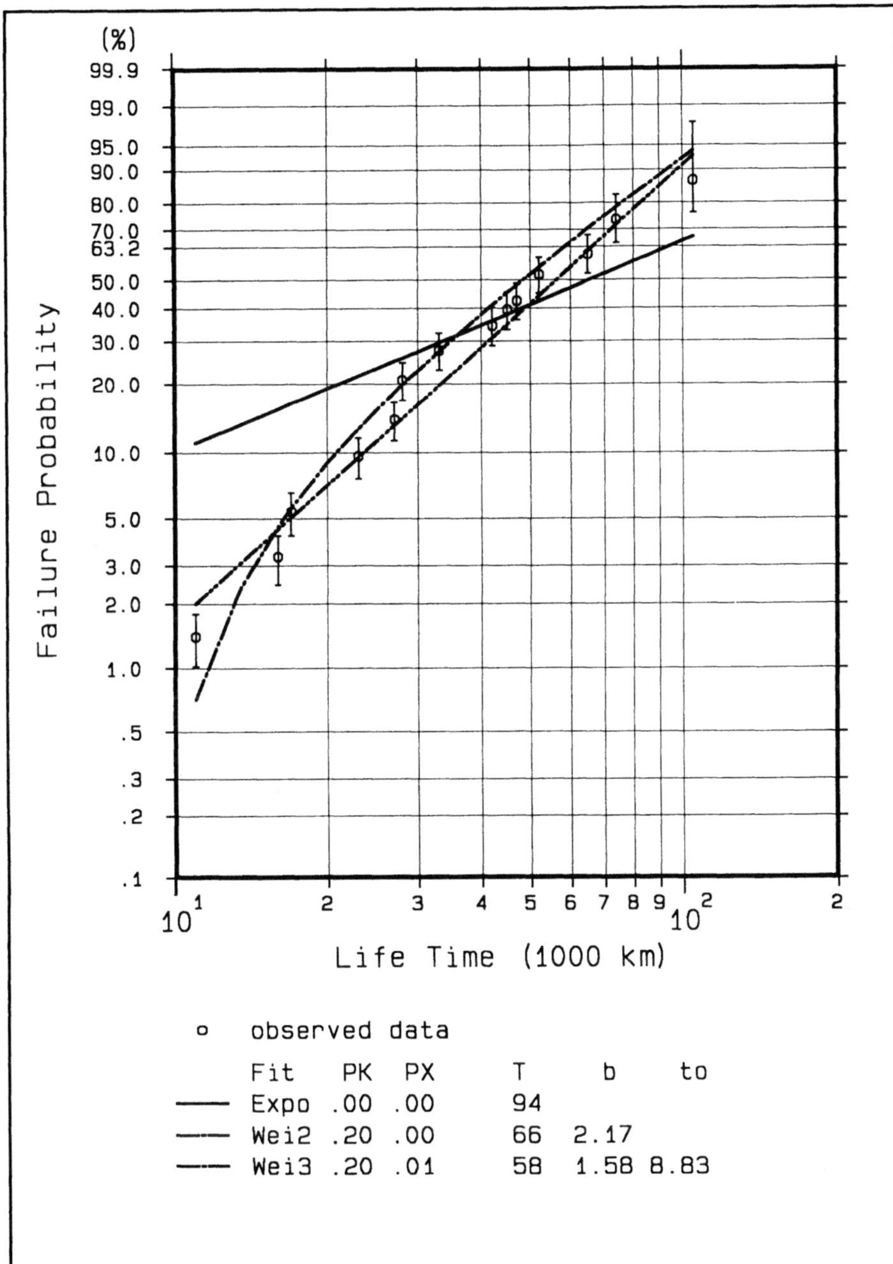

Figure 4
Graphical presentation of FDA results:
Cumulative distribution of front brake linings example [4]

BEWA - Exponential and Weibull Analysis Version 03.03.90 by R.Leicht, NUKEM

Analysis of Life Time Data File: FACB

Number of Life Times	: 56	Minimum	:	328.00
Number of Failures	: 44	Maximum	:	41018.00
Censorings	: 12	Mean	:	14568.75
Life Time Classes	: 44	Sum	:	8.16E+05
early Failures	: 0	Po	:	0.00E+00

Distribution	Fit	T	b/lower	upper	MaxDv	PKol	X^2	$P(X^2)$	<T>
Exponential	MLE	1.85E+04	1.44E+04	2.42E+04	.0934	.200	.96	.548	1.85E+04
Weibull	MLE	1.88E+04	1.17E+00		.0745	.200	1.04	.398	1.78E+04

Table 5
Results of the statistical life time analysis with FDA - power plant feed and condensate system pumps example [2]

BEWA - Exponential and Weibull Analysis Version 03.03.90 by R.Leicht, NUKEM

Analysis of Life Time Data File: FAPB

Number of Life Times	: 22	Minimum	:	31.00
Number of Failures	: 17	Maximum	:	8520.00
Censorings	: 5	Mean	:	2359.14
Life Time Classes	: 17	Sum	:	5.19E+04
early Failures	: 0	Po	:	0.00E+00

Distribution	Fit	T	b/lower	upper	MaxDv	PKol	X^2	$P(X^2)$	<T>
Exponential	MLE	3.05E+03	2.04E+03	4.79E+03	.1705	.200	1.65	.048	3.05E+03
Weibull	MLE	2.80E+03	6.73E-01		.0737	.200	.80	.678	3.68E+03

Table 6
Results of the statistical life time analysis with FDA - power plant feed and condensate system pumps example [2]

Finally, the third example, taken from car dashboard failure data [5], shows a case where no good fit at all can be achieved (fig. 7): There is a systematic effect on the data, caused by the fact that the customer claims repair of all failures during the warranty period (12 months in the present case) but hesitates later because then he has to carry the repair costs.

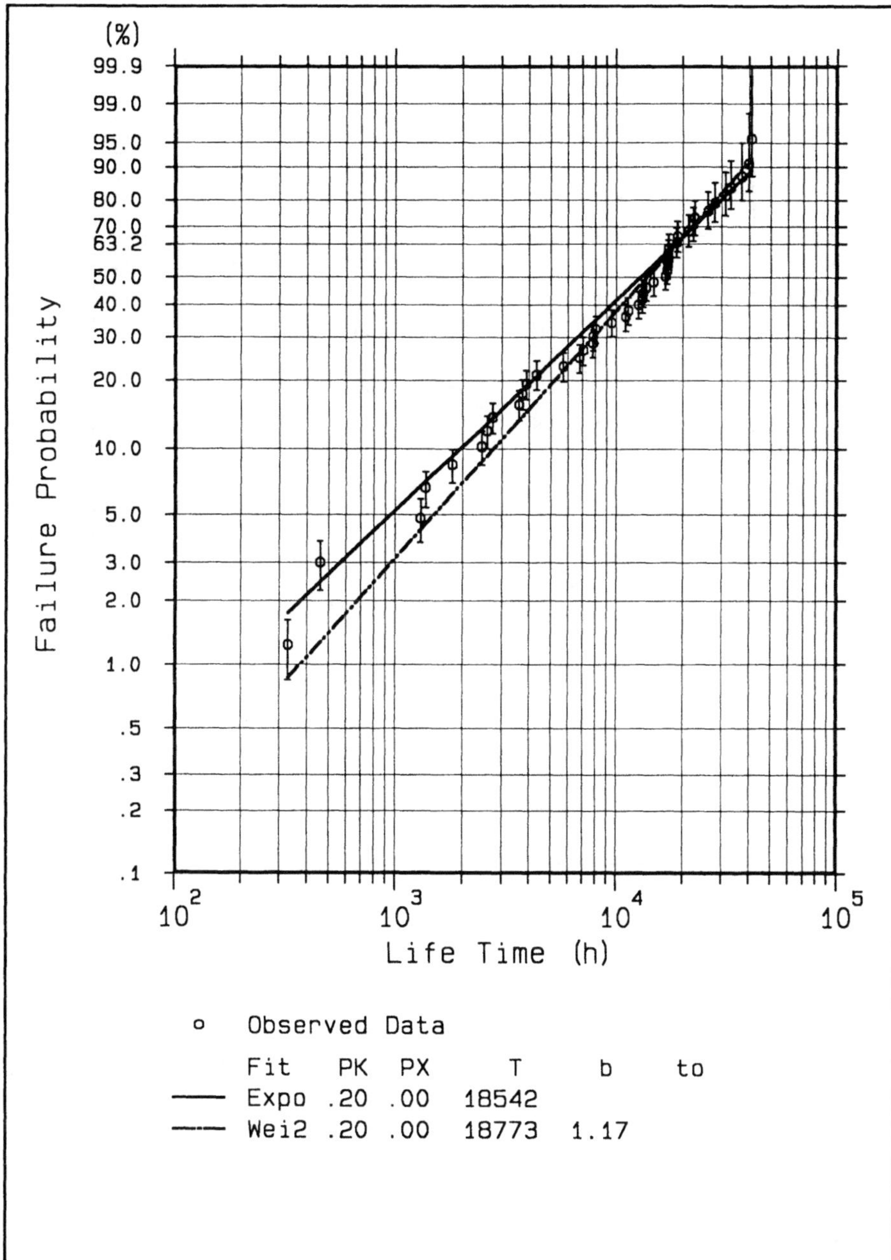

Figure 5
Graphical presentation of FDA results:
Cumulative distribution of feedwater pumps, good exponential fit [2]

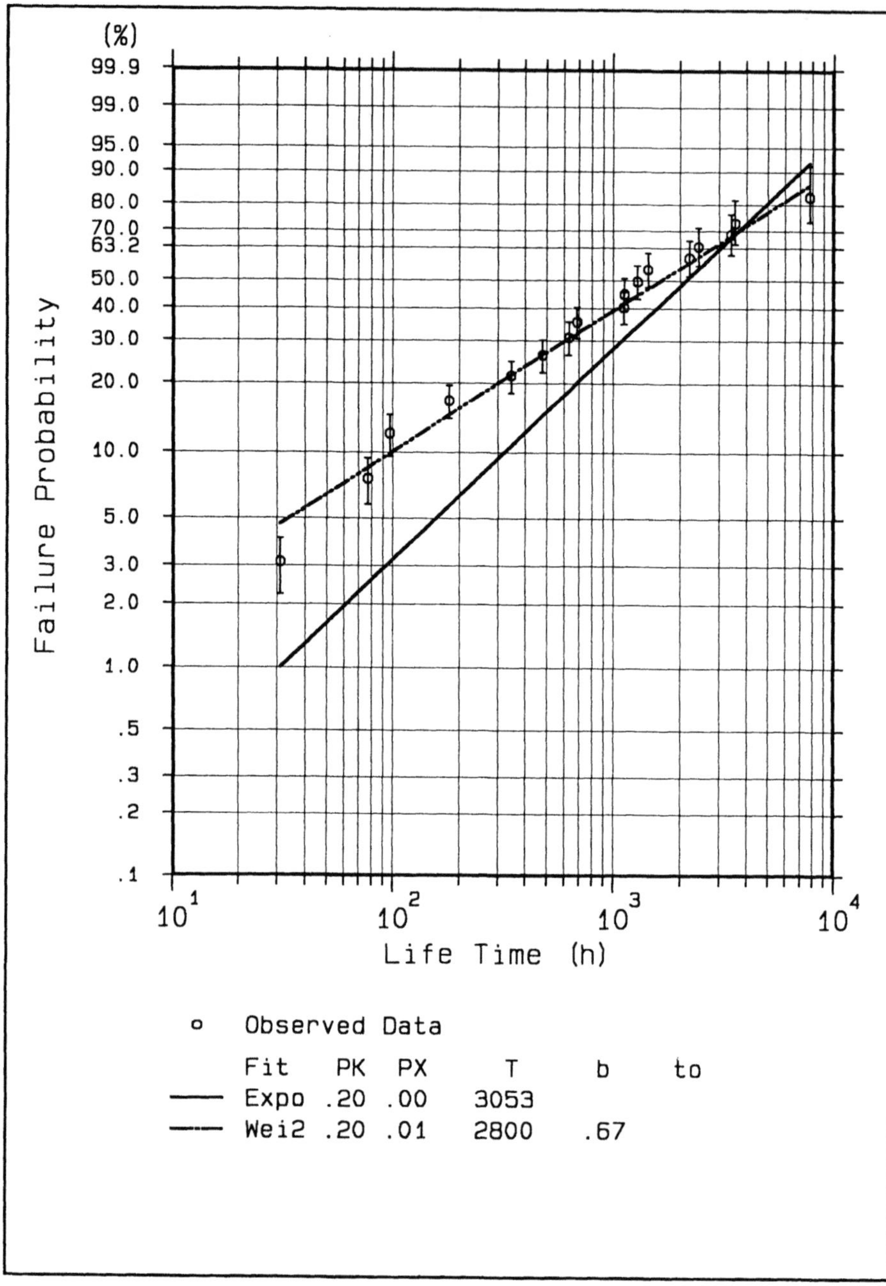

Figure 6
Graphical presentation of FDA results:
Cumulative distribution of feedwater pumps, bad exponential fit [2]

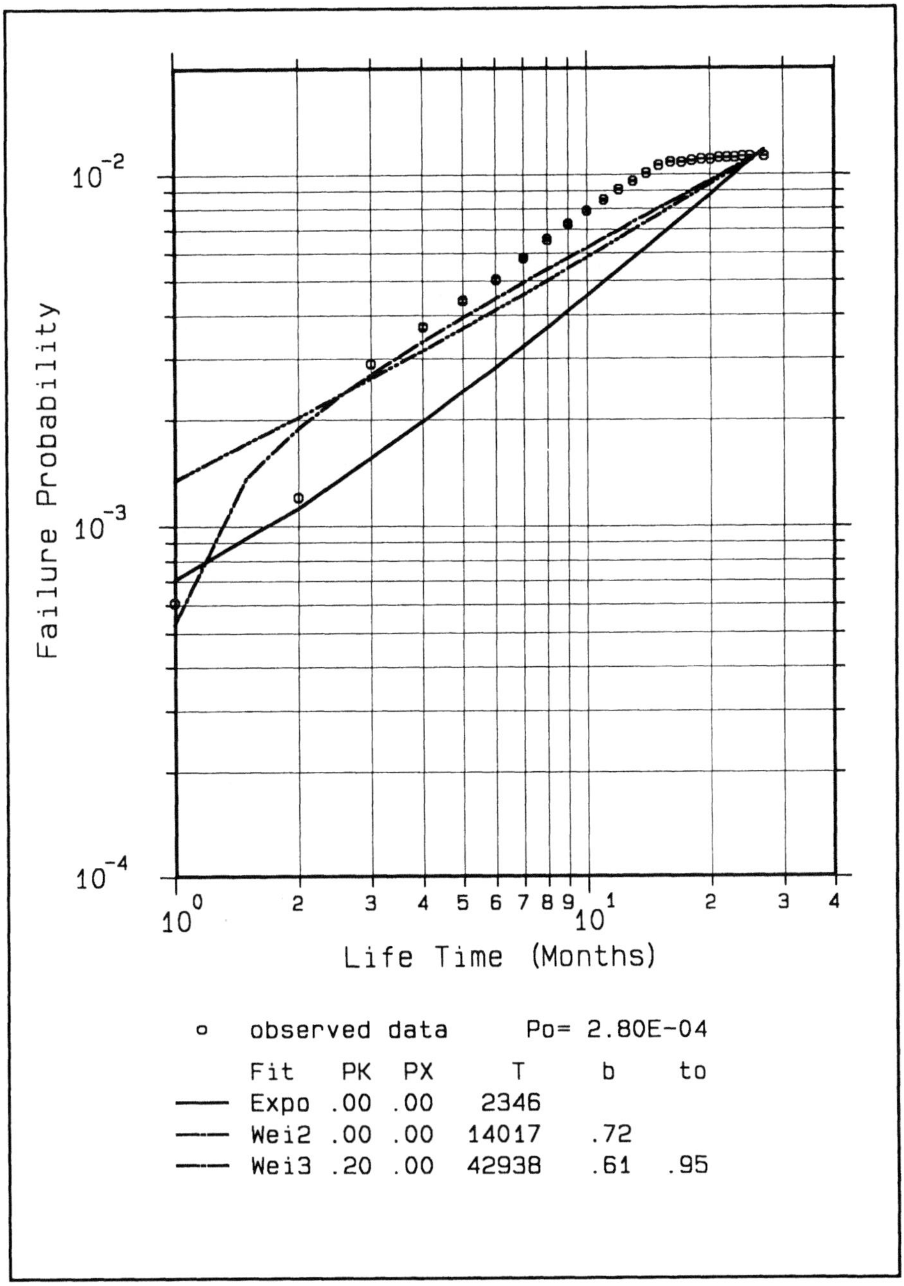

Figure 7
Graphical presentation of FDA results:
Cumulative distribution of car dashboard instruments, showing the systematic effect of incomplete data after the 12 months warranty period [5]

4. RDB - Data Bank for Reliability Data

The reliability data bank RDB stores component reliability data, such as failure rates, failure probabilities on demand, and repair times, for use in reliability or risk assessments. These data are either the result of a statistical analysis of field or test data with FDA, or are taken from other sources.

As far as available, the uncertainties related to these data are also stored, i.e. mean and variance (in the case of normal distributions) or median and error factor (in the case of lognormal distributions).

In addition to the reliability data, a characterization is stored of the homogeneous component subset for which the reliability data are applicable. This includes information about of the component type, of the underlying operating conditions, and of various aspects concerning the failure mode.

Since published data from various reliability data sources in general do not meet the same information standards, it has been found appropriate to create a separate data base with a particular structure for some of these sources.

Access to the desired information on reliability parameters for a certain component is provided in a retrieval by menu-guided iterative selection and increasing degree of specification.

The structure of RDB is simple: it consists of one single data base file. The information stored in each record can be subdivided into qualitative (descriptive) and quantitative (numerical) information, as illustrated in fig. 8.

The RDB system includes reliability data sets which have been used in reliability and risk studies over many years. Important public sources for these data were e.g. the OREDA handbook [7] or the Systems Reliability Service [8]. Other data used originate in nuclear power industry and are confidential.

R D B - Reliability Data Bank

Qualitative Information

- o description of component
- o field of application
- o operating mode
- o environmental conditions
- o maintenance
- o ...

Quantitative Information

- o observed population
- o total observed operating/calendar time
- o total number of observed demands/cycles
- o number of failures versus failure modes
- o constant failure rates versus failure mode and the corresponding confidence bounds
- o probability of failure on demand and the corresponding confidence bounds
- o mean time to repair
- o mean restoration time
- o ...

Figure 8
Information stored in the reliability data bank RDB

5. FTL - Program for Fault Tree Analysis

5.1 General

The reliability of a technical system which consists of a variety of repairable components is quantitatively evaluated in a fault tree analysis. A fault tree is the binary representation of simultaneous failure conditions which, in a system consisting of many components, lead to the failure of the system. The 'root' of the tree is the so-called top-event or top-gate which characterizes the undesired failure state.

The top gate has several inputs, which are interconnected via logical "AND" or "OR" conditions, represented by the corresponding gates in the fault tree.

Gate inputs can be other gates, or they can be so-called "basic events", which form the "leaves" of the tree and are not further developed. A basic event is e.g. the failure of a component, whose failure probability can be quantified.

The qualitative evaluation of a fault tree provides those sets of basic events whose simultaneous occurrence is both necessary and sufficient for the occurrence of the top event (top gate) (minimum cut sets). Since failure data are associated with each basic event, the frequency of occurrence of the top event and the corresponding unavailability can be quantified.

A fault tree is completely defined by the following data:

- fault tree **structure**, i.e. the set of named gates with information on the gate type (AND/OR) and the names of the related gate inputs, and the named basic events with their component type and the related failure class

- describing **texts** for gates and events

- **failure date** of the basic events

In order to manage these data optimally, a data base system is introduced which consists of the following data files:
- structure/texts file for gates and basic events
- failure data file

Because a fault tree often contains many similar components or components with the same failure behaviour, failure classes (component classes) are introduced and assigned to the particular basic events. These failure classes are stored in the failure data file.

The PC-based fault tree code FTL [9] offers the following options:

- input of fault tree data
- editing of fault tree data
- consistency check of fault tree data
- graphical evaluation and representation of a fault tree, including display on the monitor screen with editing option, output to a HPGL-plotter and output to a printer
- qualitative and quantitative evaluation according to [10]

5.2 Data Retrieval and Use in a Fault Tree Analysis

For obvious reasons it is not possible to use the brake, instruments, or pump data of the previous examples to perform a fault tree analysis. Therefore another data set is provided for the demonstration of data retrieval. The fault tree example describes the loss of cooling event of a tank with self-heating radioactive waste (HLLW). More information and data are required for the quantification of the top event probability than a generic reliability data base can offer, as can be seen from table 7. The basic events are described by the following so-called component models:

Component type 1:
These are repairable components which can fail during operation. A failure is immediately detected and repaired since the component is either selfindicating, or the failure is immediately detected by the operating personnel. The behaviour of the component is described by the failure rate and by the mean repair time MTTR.

Component type 2:
This component type is repairable, but not selfindicating. It is used to describe stand-by components which are periodically inspected or tested. Failures can only be detected by inspection or test and are immediately repaired. Thus, after an inspection these components are considered "as good as new". Their behaviour is described by a failure rate, an inspection or test interval TI, and by a mean repair time MTTR.

Component type 3:
These are repairable component which are operated cyclic or intermittend and which can fail on demand. Repair is performed immediately after the failure. The component behaviour is characterized by a failure probability on demand p and by a mean repair time MTTR.

Thus the specific conditions of the operation of a component determine the type, inspection intervals and repair times.

Class	Type	Description	
1	1	self indicating repairable rate: 1.00000 E-6/h MTTR: 8.00 h TI : 0.00 h	heat exchanger, general source: K ......: ......:
2	1	self indicating repairable rate: 0.10000 E-6/h MTTR: 8.00 h TI : 0.00 h	heat exchanger, common mode (β=0.1) source: ......: ......:
3	3	prob. of failure on demand P : 0.00100 /dem MTTR: 8.00 h TI : 0.00 h	heat exchanger, maintenance ......: 8h per year ......: ......:
4	1	self indicating repairable rate: 1.00000 E-6/h MTTR: 720.00 h TI : 0.00 h	cooling spiral, general source: as heat exchanger ......: ......:
5	1	self indicating repairable rate: 0.10000 E-6/h MTTR: 720.00 h TI : 0.00 h	cooling spirals, common mode (β=0.1) source: as heat exchanger ......: ......:
6	1	self indicating repairable rate: 50.00000 E-6/h MTTR: 8.00 h TI : 0.00 h	pump, failure during operation source: K ......: ......:
7	2	not self indicating repairable rate: 10.00000 E-6/h MTTR: 8.00 h TI : 720.00 h	pump, failure to start (R) source: K ......: ......:
8	2	not self indicating repairable rate: 50.00000 E-6/h MTTR: 8.00 h TI : 720.00 h	pump, failure during operation (R) source: K ......: ......:
9	1	self indicating repairable rate: 5.00000 E-6/h MTTR: 8.00 h TI : 0.00 h	pump, commonmode failure (β=0.1) source: ......: ......:
10	3	prob. of failure on demand P : 0.01000 /dem MTTR: 8.00 h TI : 0.00 h	pump, maintenance ......: 8 h per month ......: ......:

Table 7
Example of reliability parameters used in the reliability assessment of a HLLW tank

The top event of the example fault tree (fig. 9) describes a dangerous system state where the HLLW is boiling and the loss of liquid cannot be compensated, thus leading to a progressive concentration of the HLLW with increasing release of radioactive material. Since the boiling temperature of the HLLW is reached only 5 hours after the loss of cooling, the probability of exceeding 5 hours of unavailability must be explicitely evaluated and considered [11].

In this example system which is characterized by a high degree of redundancy of technical components, other failure contributers such as common mode failures and operator failures become important and must explicitely be included.

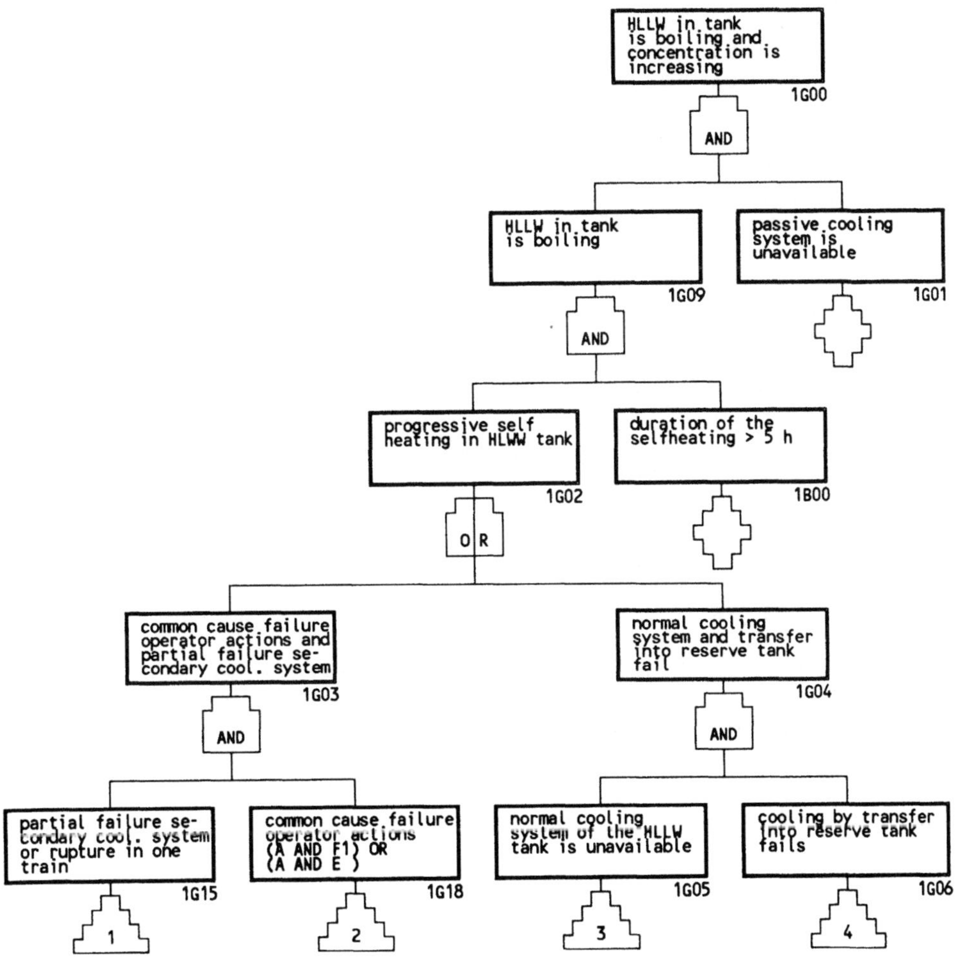

Figure 9
Top part of the example fault tree

6. Conclusion

A wide spectrum of reliability tasks has been covered in this demonstration. Computer tools appropriate for taking, storing and analyzing reliability data have been presented, and sample data and cases have been studied. Although these computer aids considerably ease the work of the reliability engineer, he must always be aware of their restrictions and limitations. Critical engineering judgement is required in order to avoid the pitfalls sometimes caused by the codes.

7. References

[1] R. Leicht and H.J. Wingender
Computer Aided Reliability and Risk Assessment
Proceedings of the 6th EuReDatA Conference, Siena
Springer Verlag, 1989

[2] A. Besi, A.G. Colombo (editors)
Proceedings of the Workshop on Reliability Data Analysis, European Reliability Databank Association, Ispra, April 1990

[3] Component Event Data Bank Handbook
Commission of the European Communities
JRC Ispra, Italy, 1984

[4] Zuverlässigkeitskontrolle bei Automobilherstellern und Lieferanten - Verfahren und Beispiele
Verband der Automobilindustrie e.V. (VDA)
Frankfurt, 1984

[5] R. Oehmke, R. Leicht und H.J. Wingender
Zuverlässigkeits- und Lebensdauerberechnung im Automobilbereich
Qualität und Zuverlässigkeit 34 (1989), pp. 673 -676

[6] FDA - a computer code for statistical analysis of failure data
NUKEM GmbH, 1990

[7] OREDA - Offshore Reliability Data Handbook
VERITAS, Hövik, Norway, 1984

[8] National Centre of Systems Reliability, NCSR Data Bank
Safety and Reliability Directorate, UK Atomic Energy Authority

[9] FTL Fault Tree Code Version 2.3 User's Manual
NUKEM GmbH, 1990

[10] VDI Richtlinie 4008, Blatt 7
Structure Function and its Application
Verein Deutscher Ingenieure, Düsseldorf

[11] A. Becker, L. Camarinopoulos
Delay Times in Fault Tree Analysis
Microelectronics and Reliability, Vol. 22, No. 4, pp 819-836, 1982

GPSR Compliance

The European Union's (EU) General Product Safety Regulation (GPSR) is a set of rules that requires consumer products to be safe and our obligations to ensure this.

If you have any concerns about our products, you can contact us on ProductSafety@springernature.com

In case Publisher is established outside the EU, the EU authorized representative is:

Springer Nature Customer Service Center GmbH
Europaplatz 3
69115 Heidelberg, Germany

Batch number: 10607320

Printed by Printforce, the Netherlands